Food Irradiation and the Chemist

Special Publication No. 86

Food Irradiation and the Chemist

The Proceedings of an International Symposium
Organized by the Food Chemistry Group of The Royal
Society of Chemistry as Part of the Annual Chemical
Congress 1990

Queen's Unversity, Belfast 10-11 April 1990

Edited by

D. E. Johnston
Queen's University, Belfast
and Department of Agriculture for Northern Ireland

M. H. Stevenson
Queen's University, Belfast
and Department of Agriculture for Northern Ireland

ROYAL
SOCIETY OF
CHEMISTRY

British Library Cataloguing in Publication Data
Food irradiation and the chemist.
  1. Food Irradiation
  I. Johnston, D. E.   II. Stevenson, M. H.   III Royal Society of Chemistry, *Food Chemistry Group*
  664.0288

ISBN 0-85186-857-6

© The Royal Society of Chemistry 1990

All Rights Reserved
No part of this book may be reproduced or transmitted in any form or by any means—graphic, electronic, including photocopying, recording, taping or information storage and retrieval systems—without written permission from The Royal Society of Chemistry

Published by The Royal Society of Chemistry
Thomas Graham House, Science Park, Cambridge CB4 4WF

Printed in Great Britain by J. W. Arrowsmith Ltd., Bristol

Preface

This book contains the proceedings of the international symposium entitled 'Food Irradiation Challenges for the Chemist' organised by the Food Chemistry Group on 10-11 April 1990 at the Queen's University of Belfast as a part of the Annual Congress of the Royal Society of Chemistry.

In view of the UK Government's recent announcement that it intends to proceed towards a change in the legislation relating to food irradiation a scientific meeting devoted to the topic is particularly relevant. Food irradiation is a topic attracting much current interest and comment, unfortunately much of it poorly informed. One of the main challenges for chemists working on irradiation is therefore the clear communication of their findings to all interested bodies. The objectives of the symposium were to bring together scientists from a range of disciplines involved in the investigation of food irradiation and through discussion to more clearly define the chemical challenges facing the food industry.

The ability to reliably distinguish between irradiated and unirradiated foods or ingredients is fundamental to providing reassurance to the public that their right to choose can be protected. Reflecting this importance, the greater part of the symposium concentrated on the scope and limitations of the more promising detection techniques currently available.

As in the medical field, it is likely that most food will be irradiated 'in pack' and this presents another series of challenges and opportunities for the chemist. The compounds produced by irradiation of food packaging materials may potentially serve as indicators that the pack has been irradiated or may migrate into the food and adversely affect its quality or safety. The most recent work in the area of food contact plastics is also presented.

The currently available techniques of dosimetry, which will be essential for the proper management and control of the industrial process, are set out and discussed.

A contribution is included on the effects of irradiation on microorganisms and the fascinating area of biochemistry associated with the irradiation induced changes.

Food irradiation has been a permitted process in the Netherlands for some time. The symposium concludes with an account of the development of commercial food irradiation in that country. This paper provides an interesting opportunity to compare approaches and to profit from the Dutch experience.

The overall perspective of food irradiation involves challenges to a wide range of individuals with a professional interest in the food industry. However it is the pressing issues of establishing scientific facts concerning detection, quality and safety of irradiated foods and communicating these clearly to the public which form the challenges to the chemist.

We would like to thank the session chairmen and all who contributed papers and discussions at the meeting. We are also grateful to the Congress Organising Committee and all involved with local arrangements whose co-operation and support contributed to the smooth running of the symposium. We thank Kathleen Dowds and her staff for secretarial support. Finally we thank the staff of the Royal Society of Chemistry for help in preparing this book.

D E Johnston
M H Stevenson

July 1990

Contents

Food Irradiation - The Challenge 1
By C.H. McMurray

Radiolytic Products of Lipids as Markers 13
for the Detection of Irradiated Meats
By W.W. Nawar, Z.R. Zhu and Y.J. Yoo

Luminescence Detection of Irradiated Foods 25
By D.C.W. Sanderson

Changes in DNA as a Possible Means of 57
Detecting Irradiated Food
By D.J. Deeble, A.W. Jabir, B.J. Parsons,
 C.J. Smith and P.Wheatley

Can ESR Spectroscopy be Used to Detect 80
Irradiated Food?
By M.H. Stevenson and R. Gray

Radiation, Micro-organisms, and Radiation 97
Resistance
By B.E.B. Moseley

Dosimetry for Food Irradiation 109
By P.H.G. Sharpe

The Effects of Ionising Radiation on 124
Additives Present in Food-contact Polymers
By D.W. Allen, A. Crowson, D.A. Leathard and
 C. Smith

Irradiation of Packaged Food 140
By D. Kilcast

Commercial Food Irradiation in Practice 153
By J.G. Leemhorst

Food Irradiation – The Challenge

C. H. McMurray

DEPARTMENT OF AGRICULTURE FOR NORTHERN IRELAND, DUNDONALD HOUSE, UPPER NEWTONARDS ROAD, BELFAST BT4 3SB, NORTHERN IRELAND

1 INTRODUCTION

There are many challenges associated with Food Irradiation. The response to the challenge depends on which challenge one listens to. Food Scientists sometimes voice the opinion that there are no challenges and may regard many of the issues associated with this food processing technique to be already addressed and indeed resolved. What is certain is that this process is one of the best researched food processing techniques in the world today. As a result the technique is now capable of taking its place alongside the many other techniques that are available to play a role in the transformation of the products of primary agriculture, fisheries and horticulture into the food we eat.

In a technological sense food irradiation must be able to demonstrate utility and in a commercial sense it will have to offer specific technological and/or economic advantage over other food processing techniques.

However food scientists have not the only voice, that of the consumer has also to be listened to. Consumers, because of the nature of the discussion surrounding the process, have expressed reservations about the technique. It will be only through examination of these issues and having such issues addressed in a scientific and rational manner that progress will be made towards acceptance for the role which the technique will play in the food chain.

The debate on the process has been stimulated to a large extent by some whose motives have been ill-defined and who often seek to confuse and confound. Nevertheless issues have been raised and these must be addressed even though the answers may not necessarily be to the liking of those who raised them in the first place.

Consumers have the following questions about such issues as:-

Is the process safe?
Will there be labelling?
Will it be possible to determine if food has been irradiated or not?
and so on.

In this introduction I have set myself the target of:-

(a) summarising the major issues; and

(b) setting the scene with respect to the presentations that come later in this symposium and which will deal in greater detail with each of the challenges that have been raised.

2 THE PROCESS

The process of food irradiation requires the use of either gamma rays, which are generated by the decay of either Cobalt-60 or Caesium-137, high voltage electrons (up to 10 MeV) or X-rays (below 5 MeV). Each of these sources of radiation causes ionisation (separation of negative and positive charge). This results in either indirect effects whereby the chemically reactive products formed from irradiated water (hydroxyl and superoxide radicals etc) are themselves chemically reactive and result in a cascade of further reactions, or direct effects resulting from chemical changes induced in molecules such as the nucleic acids present in food.

It is not generally appreciated that the radiation which triggers the chain of chemical events is part of the electromagnetic spectrum. This contains at the low energy end microwaves (communications, measurement and heating), and radio waves (communication and measurement), with increasing energy, the infra-red (heating and measurement), visible light (red to blue) region is included. At the high energy end of this spectrum occur gamma rays which are produced naturally eg emissions from the sun.

While different equipment can be used to produce the different ionising radiations, all produce the same chemical changes. The only practical differences relate to their powers of penetration and hence the dimensions and density of the food or food product capable of being irradiated.

Irradiation processing is well known in a number of industrial sectors where it makes a significant contribution (approximately 60%) to the production of sterile disposables for use in hospitals using doses of up to 30 kGy.

The very nature of the low energy used to carry out the process means that the changes induced in food have been extremely small. The direct consequence of this is that it has until now proved extremely difficult to devise unique and practical means of telling whether food has been irradiated.

This has been the most recent challenge to the Chemist and we will hear in this symposium how the Chemist (Nawar), Physical Chemist (Stevenson), Physicist (Sanderson) and Biochemist (Deeble) have responded. In addition, further chemical work by Stevenson[1] and her group has studied the formation of cyclobutanones, a group of compounds which have recently been shown to offer some potential in developing a simple detection test.

Confusion with Chernobyl

There is a world of difference between the presence of radioactivity in our food eg natural potassium-40 and the very low amounts of radionuclides resulting from the Chernobyl disaster. Such confusion often arises because of the similarity in terms eg radiation (producing radiant energy - an electric light bulb) and irradiation (being exposed to radiant energy eg sunbathing).

Purpose of Food Irradiation

Food irradiation can be used for the following processes:-

1. To enhance the safety of food by killing harmful food microorganisms.

2. To extend the shelf-life of food by killing microorganisms which cause food to deteriorate.

3. To avoid the use of harmful chemical compounds eg ethylene oxide (currently used to disinfest herbs and spices of harmful microorganisms) and methyl bromide (insect disinfestation).

4. To delay the ripening and sprouting of fruits, vegetables and fungi.

Safety of the Process

Many expert groups have examined the safety of the process. These groups have unanimously indicated that the process was safe and did not confer any enhanced toxicological, microbiological or nutritional hazard over what would be incurred by conventional food processing techniques. We only need to refer here to the UK study carried out by the independent expert group under the chairmanship of Sir Arnold Burgen.[2]

Improving Food Safety. The role of food irradiation has to be examined against a desire to enhance the safety of food. It is convenient that the technique is similarly lethal to the major pathogenic microorganisms, <u>Salmonella</u> spp, <u>Campylobacter</u> spp and <u>Listeria</u> spp which are the primary cause of food poisoning incidences in the UK at this time. Direct evidence for the rise of food poisoning is provided by the analysis of food poisoning cases regularly published by the Public Health Service Laboratories of the United Kingdom. Professor Moseley will expand on this aspect in his presentation.

Consumers have been given the impression that food irradiation is a technological fix. However like other food processing techniques which have exactly the same aim, eg heat pasteurisation of milk (reduction of pathogens and improving shelf-life) it must be used with due care and attention to processing variables which if not controlled properly can result in unsatisfactory products which no consumer would find acceptable. While it is true that the general principles of irradiation processing are understood, it is essential that each prospective application is adequately researched including pre- and post-irradiation treatment conditions in order to optimise the technique and gain the safety and technological advantages which can be delivered if the technique is used properly. It is quite clear that food irradiation cannot improve the flavours and smells associated with food deterioration. It cannot make bad food good. In fact the measure of advantage weighs towards the use of the technique as soon as possible after production and before microbiological deterioration gets under way to any significant extent.

Is the technique an excuse for poor hygiene? The answer is NO. The advantage will always be to improve hygiene as far as is practically possible. Hygiene has a finite cost but should be considered as the first option. Irradiation can then be used to enhance the safety of susceptible foods even further. However there is good evidence that food safety cannot be wholly obtained by improving hygiene alone. This constitutes in part the rationale for the Food and Drug Administration[3] in the United States giving approval for irradiation of chicken and for the British Medical Association[4] identifying the irradiation technique as the only processing technique which is likely to overcome the problem of food poisoning arising from chicken.

Concern has been expressed at the levels of toxin that will remain post-irradiation. Irradiation will kill the organisms (eg Staphylococcus) which produce toxin, so in practice less toxin will be formed compared with the unirradiated product. Furthermore it is the organism growing in the human body which is harmful and which constitutes the major threat from foodborne pathogens.

While the vegetative cells of <u>Clostridium botulinum</u> are killed, the spores of this organism remain and can grow out to form vegetative toxin producing cells post-irradiation. Considerable research has now clarified the nature of the risk and clearly demonstrated that the quantities of other bacteria remaining in the food are sufficient to suppress the growth of the botulinum organisms and inhibit the production of toxin in food post-irradiation. Food manufacturers are well used to preventing <u>Clostridium</u> <u>botulinum</u> problems in this country where there have been very few incidences of such poisoning in the human population.

<u>Extending Shelf-Life</u>. The Consumer Organisations have on occasion challenged the use of a process which extends shelf-life, claiming that such a process will have advantage only for the food processor. If anything, advantage lies with the consumer especially as purchasing trends suggest that consumers are seeking to minimise the number of purchasing trips made per week or per month. This change is also coupled with the desire to have food in a pristine condition when consumed immediately prior to the next purchasing expedition. Extension of intervals between purchases actually disadvantages the retailer who seeks to increase the frequency of purchasing in order to increase turnover of both goods and revenue.

Consumers have been erroneously led to believe that food will not deteriorate after irradiation. This would only be true if a sterilising dose was applied. Irradiation doses up to the permitted overall average maximum dose of 10 kGy will only pasteurise the food. Certain populations of microorganisms will be killed while others will remain. Indeed the nature of the final product may be different (as also occurs in milk pasteurisation) because of changes to the microbiological flora in the food. For instance, meat will normally putrify if left to naturally deteriorate whereas when irradiated meat goes off, a sour odour is generated as a result of changes in fermentation patterns.

<u>Avoiding the Use of Chemical Additives</u>. We cannot ignore the fact that the desire to remove chemical additives has enhanced the element of risk in conventional food processing techniques as far as microbiological safety is concerned. On the other hand the availability of food irradiation will permit the reduction in the use of compounds such as ethylene oxide and methyl bromide, compounds which Governments are removing from the treatment of food, because of either the risk of cancer associated with the residues that remain or additionally because they are toxic to humans involved in the treatment process.

<u>Delaying Ripening and Sprouting</u>. Irradiation alone offers significant advantages in the storage of fruit and vegetables. Low dose irradiation (0.1 kGy)

will inhibit the sprouting of potatoes stored at 15°C. Such a treatment obviates the need for using chemical sprout suppressants such as Tecnazene. The technique can be used as an alternative to cold storage which is not only more expensive than irradiation but also results in the accumulation of reducing sugars which when the potatoes are deep fried (chipped or crisped) will result in an enhanced and undesirable browning reaction.

A further role is the use of the technique to delay ripening in tropical fruits. In order for such fruit to arrive in our homes in an edible condition, it is necessary for the fruit to be packed in an unripened state. The fruit ripens during transport or in suitable storage conditions in this country. The flavour of such fruit bears little relationship to the flavour encountered during the natural ripening process. Irradiation allows the fruit to ripen naturally. The ripening process is then arrested, hence the fruit can be consumed not only with its natural flavour and aroma characteristics but also with its physical properties intact. Again it is readily demonstrated that optimum dose and storage regimes must be used in order to gain full advantage of the technique.

Packaging

A further significant advantage for the consumer in terms of safety lies in the fact that food can be irradiated after packaging. Such a procedure prevents bacteriological cross-contamination occurring during both retailing and transport and contamination will not occur until the package is opened immediately prior to cooking or consumption. Such an advantage can only be claimed if the irradiation process alters neither the nature of the packaging material nor induces chemical changes so that residues could be formed which would contaminate the food coming directly in contact with the packaging material. This aspect will be addressed by Allen and Kilcast later in the proceedings.

Combination Treatments

The treatment of food by irradiation has assumed that the process will be used on its own. This is unlikely to be the case especially with perishable products. It is more likely that significant advantages will be gained by combining the process with chilling and/or freezing. Further opportunities are on the horizon for using the technique in combination with other food processing techniques. This is the subject of active research in Belfast.[5] Furthermore in using such combination treatments it may be possible to achieve the same safety or technological advantages using lower doses of irradiation than would normally be the case.

Like all food processing techniques it is essential that the interactions of the different variables are fully explored, especially with respect to organoleptic characteristics so that the boundary between advantage and disadvantage can be determined.

There is certainly no mileage in introducing a technique which would confer unexpected properties to food thus causing refusal on the grounds of unfulfilled expectations.

3 INTERNATIONAL DIMENSIONS AND CONSUMER ACCEPTANCE

The use of the technique has been under consideration by Governments for some considerable time. Each country has adopted its own unique approach to the introduction and licensing of the technique. This approach has resulted in a patchwork of clearances and applications. The reasons for this have been complex and are beyond the scope of this short introduction. Individual responses have been dictated by the perceived needs on one hand and the counter pressures that have been brought to bear. While there has been a general decision to move forward,[6] this has been modified on the other hand by a reluctance of consumers to adopt a technique which is still perceived as being novel. There is no doubt that pressure to adopt the technique is more acute in hot humid (tropical) climates where the problems of both enteric disease and food wastage are more critical. However failure to adopt the process in developed countries (usually moderate climates) has inhibited the uptake in those regions (less well developed) where significant advantage could be gained.

Consumers are being asked in this country to accept a theoretical proposition. While there have been surveys showing considerable uncertainties about irradiated food they are only assessing opinion against a perception of the technique and what it will do. In these surveys 25% of those interviewed have said they will eat irradiated products against 50% who are undecided. This contrasts with what has happened in those countries where proper acceptance trials have been conducted and in which consumers were offered a free choice between irradiated and unirradiated products. In such trials there has been a majority choosing the irradiated product - on the grounds of both improved flavour and quality standards.

Such tests have been conducted in the USA (papaya), Thailand (fermented sausage), Poland (onions) and France (strawberries).

4 CONTROL

The World Health Organisation has developed control regimes for the use of the process.[7] In the United Kingdom the Government is currently amending the legislation so that the existing prohibition on the use of the technique will be removed. Further work has to be carried out to develop appropriate regulations and a licensing regime both of which have to be in place before the technique can be used commercially. However thought has already been given to what will be required in these regulations[8] and a framework will come forward for consultation and debate in due course.

<u>Labelling</u>

There is no disagreement that food that has been irradiated should be adequately labelled in order to give consumers a choice in what they will eat. While a logo has been suggested as being an adequate identifier for irradiated food, the view in the United Kingdom is that this is inadequate and a label incorporating a term such as "Treated with Ionising Radiation" should be used to clearly differentiate the product.

<u>Dosimetry</u>

It is currently the intention that adequate documentation will accompany irradiated food through the food chain. The primary control mechanism will be at the irradiation plant. A key element of this is the proper use of dosimetry to ensure that the technique is optimally used. Dr Sharpe explores the properties of the various dosimetry systems in his paper. It will be important to have a dosimetry system that will be traceable to a national standard.

Dosimetry is a very effective control on the proper use of the technique. Nevertheless consumers have been demanding that, in order to enforce the labelling requirements, independent tests should be available to check if food has been irradiated. It must be said that not all labelling requirements for our currently available food are capable of being policed. A prime example is organic foods whose labelling cannot be enforced through detection. These foods normally sell at a premium and as a result could very well attract sharp practice and unscrupulous traders willing to exploit unwary and unprepared consumers.

Detection Tests

The papers in this symposium have already been referred to and illustrate the progress that is being made in developing suitable detection tests. In order to set the scene it is necessary to develop a set of criteria against which the various tests can be compared and to assess their utility and potential for application in a range of food stuffs.

The following questions should be addressed by any prospective test methodology.

1. To what range of foods does the test apply?

2. Is the test capable of being used either as a quantitative or a qualitative technique?

3. Is the test specific for irradiation or will other processes cause similar changes?

4. What is the test's lower limit of detection and does its working range cover doses likely to be encountered in the industry?

5. What is the selectivity of the test in detecting irradiated ingredients in admixture with unirradiated foods?

6. Do the test results depend upon irradiation process variables such as dose rate, temperature of the treatment etc?

7. Do the test results depend upon pre- or post-irradiation storage variables of the food such as physiological age, temperature etc?

8. What are the accuracy and reproducibility of the test?

9. Does the test require a single measurement or does interpretation require a profile of measurements?

In order to answer these and many other questions it is essential to apply the rules of sound analytical chemistry and to subject the resulting analytical protocols to the full rigour of national and international verification involving ring tests and blind trials. Such procedures will be necessary for the tests to be accepted in a Court of Law and for them to have any significant role in the regulation of both national and international trade. Such trials are currently underway for some detection tests and the results will be published in due course.

In this respect the true challenge is for the Chemist. It is only if complete rigour is applied that credible progress will continue to be made.

5 CONCLUSIONS

Chemists have a significant role to play in understanding the basis and application of food irradiation and especially in explaining to a sceptical public what the process entails and how it can be properly used to advantage to enhance the food we eat.

REFERENCES

1. M.H. Stevenson, A.V.J. Crone and J.T.G. Hamilton, Nature, 1990, 344, 202.

2. Advisory Committee on Irradiated and Novel Foods (ACINF), 1986. Report on the safety and wholesomeness of irradiated food, HMSO, London.

3. FDA, "Irradiation in the Production, Processing and Handling of Food; Final Rule, 21CFR Part 179", Federal Register, 1990, 55, No. 85.

4. British Medical Association, 1989. "Infection Control, the British Medical Association Guide". p. 122. Edward Arnold, London.

5. M.F. Patterson and I.R. Grant, 1990. Proceedings of Irradiation Combination Treatments Conference, March, London.

6. FAO/IAEA/WHO/ITC-UNCTAD/GATT, 1988. International Conference on the Acceptance, Control of, and Trade in Irradiated Food. WHO, Rome.

7. General Standard for Irradiated Foods and Recommended international code of practice for the operation of radiation facilities used for the treatment of foods, 1984. <u>Codex Alimentarius Commission</u>, <u>15</u>.

8. Working Party of Officials of the Agriculture and Health Departments, 1989. Report on the introduction of food safety in the UK. MAFF, London.

Radiolytic Products of Lipids as Markers for the Detection of Irradiated Meats

W. W. Nawar, Z. R. Zhu and Y. J. Yoo

DEPARTMENT OF FOOD SCIENCE, UNIVERSITY OF MASSACHUSETTS, AMHERST, MA 01003, USA

1 INTRODUCTION

The possibility of using lipid-derived hydrocarbons as indicators of irradiation in food was first proposed by us in 1970.[1] The rationale behind this approach was based on the observation that, when exposed to high-energy radiation, fatty acids undergo preferential cleavage in the ester carbonyl region giving rise to certain "key compounds" in relatively large quantities.[2] Two hydrocarbons from each fatty acid were particularly attractive. One has a carbon atom less than the parent fatty acid and results from cleavage at the carbon-carbon bond alpha to the carbonyl group; the other has two carbon atoms less and one extra double bond, and results from cleavage beta to the carbonyl. This phenomenon, originally observed in experiments with model systems of fatty acids, esters and triacylglycerols[2,3] was clearly detectable in the radiolytic products of beef and pork fats,[4] fish oil,[5] and vegetable oils.[6]

In our original publication we reported on the feasibility of detecting irradiation in pork meat at doses between 1 and 60 kGy by analysis of the six "key hydrocarbons", i.e. heptadecene, hexadecadiene, pentadecane, tetradecene, heptadecane and hexadecene, typically produced from the three major fatty acids of pork fat, i.e. oleic, palmitic and stearic. A linear relationship between radiation dose and each of these compounds was demonstrated, and the presence of air or moisture was seen to have no significant effect on the quantitative pattern.

The purpose of the present study was to examine the reliability, sensitivity and practicality of this method when applied to meats and poultry.

2 EXPERIMENTAL PROCEDURES

Materials and Irradiation Treatment

Unfrozen, deboned beef, pork and chicken breasts (skinned), were purchased locally, ground and portions, 100-150 g each, packaged in polyethylene bags, sealed in air and kept frozen until irradiation. Samples were irradiated at ambient temperature with Cobalt-60 at the University of Lowell Irradiation Laboratory. After irradiation, the samples were packed in dry ice, brought back to our laboratory, a two and one half hour journey by car and kept at a temperature of -20°C until analyzed. Fatty acid composition of the meat samples is given in Table 1.

Analysis of Volatiles

The lipids were extracted from the meat samples with pentane/isopropanol (3/2, v/v), dried and subjected to high-vacuum cold-finger distillation as described previously.[7] The volatiles were then collected and fractionated with silica gel into polar and nonpolar fractions. Based on previous studies, we focused on the nonpolar fractions. A Supelcowax 10 fused silica capillary column (30m x 0.32mm i.d.) was used for gas chromatographic separation. Quantitative measurements were made with appropriate internal standards (tridecene and 3-undecanone).

3 RESULTS AND DISCUSSION

Typical gas chromatographic analyses of the nonpolar volatiles from chicken, beef and pork meats, irradiated at 2 kGy, are shown in Figure 1. It can be seen that the six radiolytic hydrocarbons mentioned above are easily detected. In the case of chicken, the four hydrocarbons, 15:1 and 14:2, and 17:2 and 16:3, produced from the 16:1 and 18:2 fatty acids, respectively, could also be employed as indicators of irradiation since the content of these two parent acids in chicken fat is significant.

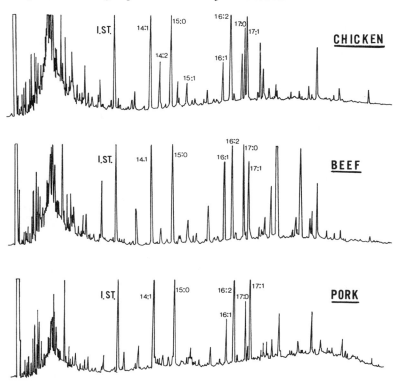

Figure 1   Gas chromatographic analyses of the volatile hydrocarbons from meats irradited at 2 kGy

Sensitivity and Repeatability

   The analyses of the chicken samples from separate irradiations were combined and statistically analyzed. In an effort towards a conservative approach, possible differences from such variables as accuracy of irradiation dose, fatty acid composition, storage time, GC detectability, analysts' experience, etc., were intentionally neglected. The results are shown in Figure 2 and Tables 2 and 3. Although the linear relationship between radiolytic products and dose is seen for all the hydrocarbons tested, it is evident that hexadecadiene, heptadecene and tetradecene are the most suitable indicators of irradiation. These compounds satisfy the

Table 1 Fatty Acid Composition (%)

Fatty Acid	Chicken	Pork	Beef
14:0	0.8	1.5	2.9
16:0	24.6	23.7	24.4
16:1	7.6	2.8	3.3
18:0	5.8	13.5	16.5
18:1	38.6	43.8	41.3
18:2	17.3	8.4	2.4
18:3	0.6		

two conditions required for an effective marker, i.e. absence, or low level, in the unirradiated control (A_0) and a high sensitivity (A_1) to irradiation. The relative error based on 95% confidence level was approximately 10%.

Standard Curves

All data obtained from the analyses of radiolytic hydrocarbons in the irradiated chicken, beef and pork, were pooled to prepare standard curves from the 16:2, 17:1 and 14:1 hydrocarbons which, as indicated above appear to be the most practical indicators of irradiation. Since the amount of each hydrocarbon produced depends on the concentration of its substrate fatty acid in the food, standard curves were also prepared by plotting amounts of radiolytic hydrocarbons versus amounts of substrate fatty acids (Figures 3-5).

Blind Tests

Experiments were designed to assess the ability of the technique to determine (a) whether a meat sample of unknown history has been irradiated, and (b) the approximate dose received.

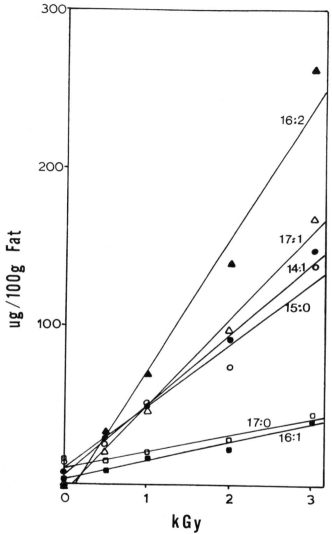

Figure 2 Effect of dose on the radiolytic hydrocarbons in chicken

Table 2 Quantitative Analysis of the Hydrocarbons Produced in Chicken (μg/100 g fat) by Irradiation at Different Doses

Hydrocarbon	0 kGy	0.5 kGy	1 kGy	2 kGy	3 kGy
15:0	14.1+1.6[1] (22)[2]	24.6+2.6 (11)	52.3+4.4 (24)	75.2+7.1 (13)	139.1+10.6 (7)
14:1	7.9+1.9 (26)	27.8+1.1 (13)	50.0+3.2 (32)	92.1+7.8 (15)	148.8+4.7 (6)
15:1	0 (22)	3.2+0.8 (13)	9.2+1.2 (26)	18.0+1.6 (14)	35.3+2.4 (6)
14:2	0 (22)	5.7+0.8 (13)	15.6+1.8 (27)	28.8+2.4 (14)	63.6+4.1 (6)
17:0	15.3+1.7 (24)	15.5+3.0 (12)	20.4+2.1 (23)	26.7+2.1 (19)	46.0+5.4 (6)
16:1	2.6+1.1 (25)	9.0+1.6 (15)	14.4+1.6 (26)	23.0+1.8 (15)	41.1+4.4 (7)
17:1	0.2+0.3 (22)	21.5+2.3 (14)	46.0+4.4 (26)	97.8+7.4 (12)	168.2+22.1 (6)
16:2	0 (27)	31.9+3.5 (12)	69.3+7.2 (26)	140.5+10.8 (13)	263.0+20.3 (7)
17:2	0 (20)	8.9+1.1 (13)	19.9+2.2 (21)	36.8+4.6 (14)	78.2+10.5 (6)
16:3	0 (20)	6.6+1.3 (13)	17.2+3.6 (21)	31.6+2.7 (14)	100.7+12.8 (6)

[1] 95% confidence level

[2] No. of samples given in parenthesis

Table 3. Slopes (A_1), y intercepts (A_0), and correlation coefficients of curves representing dose vs radiolytic hydrocarbons.

Hydrocarbon	A_0	A_1	Correlation Coefficient
15:0	8.4	40.6	0.9823
14:1	4.8	46.6	0.9975
15:1	-1.9	11.6	0.9867
14:2	-3.9	20.5	0.9760
17:0	11.7	10.1	0.9536
16:1	2.1	12.3	0.9893
17:1	-5.7	55.7	0.9956
16:2	-10.9	86.1	0.9946
17:2	-3.9	25.1	0.9796
16:3	-9.8	31.5	0.9342

Samples of chicken, beef and pork were irradiated at 0.25, 0.5, 1 and 2 kGy at ambient temperature and stored at -20°C. The research group for this test consisted of a "Director" who was the only person with access to the code listing of the samples and doses received, an "Assistant Director" who was made aware that only the above four doses were applied but was blind to the specific dose each sample received, and four "Analysts" who conducted random analyses of the irradiated samples and computed the doses received from the standard curves of the 17:1, 16:2 and 14:1 hydrocarbons. The doses based on these three indicators in each sample were then averaged and recorded as "Calculated Averages". Being aware that only 0, .25, .5, 1 and 2 kGy were applied, the Assistant Director computed the Estimated Dose for each sample by selecting the nearest dose to the Calculated Average (Table 4).

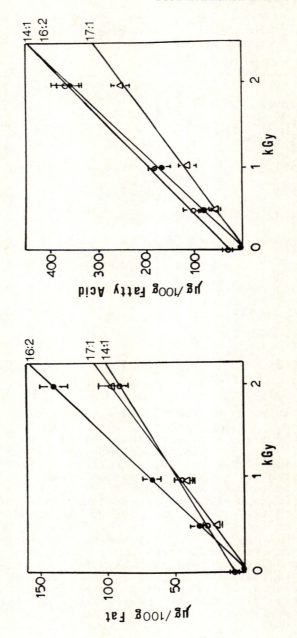

Figure 3 Standard calibration curves for chicken

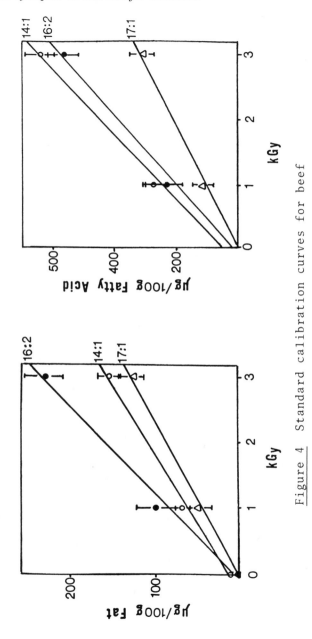

Figure 4 Standard calibration curves for beef

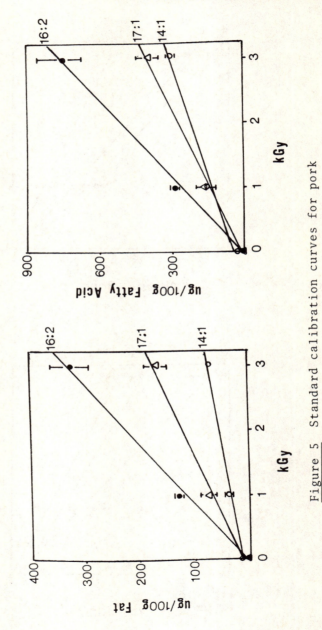

Figure 5 Standard calibration curves for pork

Table 4 Example[1] of Dose Estimation in Blind Tests for Irradiation in Chicken (kGy)

Sample Code	Calculated From 14:1	16:2	17:1	Calctd. Ave.	Estmtd.	Actual
C1	2.22 ± .20[2]	1.99 ± .15	1.95 ± .15	2.05	2	2
C2	0.14 ± .04	0.23 ± .02	0.29 ± .02	.22	.25	.25
C3	0.03 ± .02	0.03 ± .00	0.09 ± .01	.05	0	0
C4	0.73 ± .08	0.52 ± .04	0.64 ± .05	.63	.5	.5
C5	1.26 ± .12	0.88 ± .07	0.97 ± .07	1.01	1	1
C6	0.42 ± .05	0.33 ± .03	0.46 ± .04	.40	.5	.25

[1] Of a total of 31 chicken meat samples only 6 are shown in this table as example

[2] 95% confidence level

Final decisions (irradiated/not irradiated) were made by comparing the estimated values with those actually received. Since the minimum dose applied in this experiment was 0.25 kGy and considering the detection limits of gas chromatography, the following criterion was used:

A sample with a Calculated Dose higher than 0.125 kGy was judged to be Irradiated. A Calculated Dose of or below 0.08 kGy indicated that the sample was Not Irradiated. Calculated doses in the range 0.08 - 0.125 were deemed Difficult to Judge.

For a total of 81 meat samples, the technique gave a 100% correct determinations of whether irradiation has been applied. For the chicken, only 2 samples irradiated at 0.25 kGy were misjudged as receiving 0.5 kGy, and 2 others irradiated at 0.5 kGy misjudged as receiving 1 kGy. For beef, only one sample was misjudged as receiving 2 instead of 1 kGy. For pork, 4 samples were misjudged as receiving 0.5 instead of 1, and 3 samples as receiving 1 instead of 2 kGy.

The data presented here clearly indicates that the proposed method can be reliably employed for the detection of irradiation treatment in meats. It is also evident that while we do not suggest the use of food as a "radiation dosimeter", this technique is capable of providing a close approximation of the dose received. We plan to co-ordinate a collaborative program in which several laboratories in and outside the U.S.A. would participate in further testing and standardization of this method.

ACKNOWLEDGMENT

This work was supported in part by a grant from the USDA Food Safety and Inspection Service and University of Massachusetts Agricultural Experiment Station Project No. 586. The authors thank Donald Derr and Regina Whiteman.

REFERENCES

1. W.W. Nawar and J.J. Balboni, J. Assoc. Off. Anal. Chem., 1970, 53, 726.

2. M.F. Dubravcic and W.W. Nawar, J. Amer. Oil Chem. Soc., 1968, 45, 656.

3. P.R. LeTellier and W.W. Nawar, J. Agr. Food Chem., 1972, 20, 129.

4. J.R. Champagne and W.W. Nawar, J. Food Sci., 1969, 34, 335.

5. M.F. Dubravcic and W.W. Nawar, J. Amer. Oil Chem. Soc., 1969, 17, 639.

6. J.P. Kavalam and W.W. Nawar, J. Amer. Oil Chem. Soc., 1969, 46, 387.

7. W.W. Nawar, J.R. Champagne, M.F. Dubravcic and P.R. LeTellier, J. Agr. Food Chem., 1969, 17, 645.

Luminescence Detection of Irradiated Foods

D. C. W. Sanderson
SCOTTISH UNIVERSITIES RESEARCH AND REACTOR CENTRE, EAST
KILBRIDE G75 0QU, UK

1 INTRODUCTION

The need for forensic tests to identify irradiated foods has been widely recognised at a time of growing international trade in such products and impending changes in UK and EEC legislation to control the process. This paper outlines the requirements for and of such tests, and discusses recent developments in luminescence approaches aimed at meeting the needs of public analysts, retailers and consumers. Detecting whether or not food has been irradiated, and if so to what dose, is one of the challenges which food irradiation poses to the scientist.

2 USES AND CONTROLS ON FOOD IRRADIATION: THE CURRENT POSITION

The applications of ionising radiation to foods have been under study almost since the discovery of radioactivity. The primary interactions between gamma rays or electron beams and matter are well known and understood, leading to the production of energetic secondary electron-ion pairs at total concentrations which are proportional to the radiation dose absorbed in the medium. Process conditions, such as sample temperature or form have little effect on this primary ionisation process. At a dose of 10 kGy for example the primary electron-ion pair concentration is of the order of 100 ppm in typical food matrices. The heat equivalent of this energy transfer results in sample warming of less than 3 degrees centigrade.

Subsequent in-situ formation, diffusion and chemical reaction of radiolytic free radicals and excited molecules result in quite remarkable biological changes in foods and the pests and micro-organisms which reside on them.[1,2,3] These secondary and tertiary radiation effects are highly sensitive to process variables such as dose-rate, temperature, oxygen and moisture content, and the composition of the food itself. In addition to dose level - which determines the statistical proximity of free charge carriers within the medium - process conditions dictate the sizes and types of chemical and biochemical structures which are subject to effective stochastic alteration by radiation treatment.

The microbiological effects and impact on minor constituents (eg vitamins) are thus sensitive to process conditions. Not all foods are well suited to irradiation under standard conditions (ie ambient temperature in air); rancidity of fats, adverse flavour changes and reduced vitamin content are possible side effects at high doses. Nevertheless in many cases practical benefits such as shelf life extension, inhibition or, in some cases elimination, of parasites, pests, and live vegetative bacteria, including some pathogens, can be achieved at dose levels which do not produce major chemical or structural changes to the products. In contrast with other food preservation technologies such as heating, freezing, canning or drying, irradiation produces minimal chemical or physical change to achieve its effects and operates throughout the volume of the food rather than spreading from the surface. Providing a detailed account of the minor chemical changes which take place in complex foodstuffs exposed to doses below 10 kGy certainly provides a challenge to the food chemist.

The wholesomeness and safety of foods irradiated below 10 kGy have been endorsed by a number of international or national committees,[4-9] following reviews of available data on the microbiological, toxicological, nutritional and radiological implications. Within this body of opinion there is an overall consensus that irradiation of foods to average doses up to 10 kGy presents no special hazards, subject to appropriate dose control and the adoption of good manufacturing practice. The potential impacts on post-harvest losses in developing countries, and on alleviating food-borne diseases have not yet been explored practically on a large scale. However there are obvious possibilities in these areas for the beneficial use of atomic and nuclear applied science. The economics of establishing plants based on cobalt-60 or machine sources may limit widespread commercial development. Nevertheless over 30 countries already permit and practise some form of food irradiation,[10] the most common applications being in disinfestation and reduction in microbial load from spices and in treatment of shellfish and other tropical products to eliminate potential pathogens. Table 1 illustrates some of the potential applications, covering a dose range from 50 Gy to 7 kGy.

Regulatory frameworks vary considerably from country to country; ranging from total bans to virtually unrestricted application. Where irradiation is permitted regulations are needed to license the plant and any associated radioactive materials, to ensure radiation safety, environmental security and general health and safety during plant operation, and to ensure safe disposal of any hazardous materials at the end of the operation. Machine sources of radiation share some of these constraints. In this respect food irradiation presents the same regulatory challenges as the use of gamma or electron beams for materials processing or medical product sterilisation.

Additionally the acceptance of irradiated food, particularly in countries where consumer choice and preferences are important political, social and commercial considerations, depends on enforceable

TABLE 1 MAIN POTENTIAL USES OF RADIATION FOR FOODS[11]

Purpose	Dose / kGy
Inhibition of sprouting	0.05-0.15
Delaying Ripening	0.5 - 1
Insect Disinfestation	0.15-0.5
Shelf life extension	1 - 3
Elimination of spoilage and pathogenic organisms	1 - 7
Improving organoleptic qualities	2 - 7

controls to ensure adoption of good radiation practices. Adequate dose control, proper physical segregation of treated and untreated products, assuring the microbiological quality of products for treatment, and adherence to labelling rules fall within this category. These are areas for national regulation within international guidelines; they are also fields in which the technology produces yet more challenges for the analytical scientist. Those countries following the FAO/WHO Codex accept a need for labelling irradiated foods as the basis for informed consumer choice and ultimate acceptance of the process.

The UK position at the time of writing, based on statutory instruments from 1967 and 1972 [12,13] prohibits the production, importation or trade in irradiated food for human consumption, with the exception of the production of sterile diets (irradiated at >25 kGy) for certain patients undergoing medical treatment, and the use of low dose x-ray equipment in food sorting and testing. However in 1986 the UK Advisory Committee on Irradiated and Novel Foods (ACINF) reported that there were no grounds, on the basis of safety, for maintaining the present ban, and recommended general clearance of food irradiation up to 10 kGy average doses[6] in accordance with Codex Alimentarius recommendations, using gamma ray or electron beam sources of less than 10 MeV. This view has been reaffirmed following public consultation,[7] and forms the basis for proposed legal and regulatory changes currently on the UK parliamentary agenda. A vocal anti-radiation movement has also emerged in the UK, challenging most of the axioms behind the proposed changes, and lobbying vigorously to oppose them. The outcome is still awaited.

The EC position is also likely to change. At present diverse national regulations and corresponding trade restrictions apply. For example the Netherlands, Belgium and France have active commercial irradiation programmes covering several products, and subject to differing rules regarding permitted foodstuffs, microbiological pre-conditions and labelling. Other member states, for example West Germany, do not accept irradiated imports from these countries and have not expressed a willingness to adopt the process. The European Commission has proposed a system of product by product approval, outlined in Table 2, to the European parliament. At the time of writing only the use of irradiation to treat herbs and spices has been approved through this route. The effect that this lack of uniformity will have on the removal of trade barriers in

TABLE 2 1989 EEC PROPOSALS[14]

Foods	Overall Dose / kGy
1. Fruits	up to 2.0
2. Vegetables	up to 1.0
3. Cereals	up to 1.0
4. Starchy Tubers	up to 0.2
5. Spices	up to 10
6. Fish and Shellfish	up to 3.0
7. Fresh Meat	up to 2.0
8. Poultry	up to 7.0

1992, and vice-versa is still uncertain. Clearly food irradiation presents legislative and diplomatic challenges in this respect.

The technical merits of regulated irradiation are clearly demonstrable for certain products in comparison with alternative processes. For example the use of irradiation to control the microbiological quality of herbs and spices is arguably preferable to, and certainly less damaging than, any alternative method - including chemical, heat or steam treatments. Nevertheless there is considerable public and consumer lobby antipathy to the process and concern about possible abuses. A number of cases of abuse have been documented, usually involving reduction of unacceptably high levels of microbiological contamination'[14] coupled to fraudulent concealment of the irradiation treatment.

Reliable forensic tests to identify irradiated foods are needed to regulate the process, to provide commercial organisations with the means to verify the authenticity of claims prior to food purchase, and to provide assurances to consumers that labelling regulations, or import prohibitions can be enforced.

3 REQUIREMENTS OF TESTS

A primary requirement of a forensic test is that it should identify the preceding irradiation event unambiguously. Many radiation induced chemical or microbiological changes to foods can also be produced by other means and thus carry additional connotations which are unrelated to radiation exposure.

Beyond this there is an important distinction between **qualitative** tests which simply denote irradiation and **quantitative** tests which provide additional dosimetric information.

The need for qualitative tests applies equally to the maintenance of prohibitions on the process as to the enforcement of labelling, or pre-purchase validation in those countries which permit it. Good qualitative tests are robust to false negative (failure to detect irradiated samples) and false positive (mis-identification of unirradiated samples as irradiated) results. For forensic purposes the requirements are for changes which can be detected reliably from all irradiated samples of a given type, and which cannot be produced by treatments other than exposure to ionising radiation within the range of possible food processing doses.

Qualitative indicators which can be produced by other means than radiation, or which work under restricted circumstances, will be of reduced value. They may however be of some use in combination with other information, providing that their failure mode is well understood and can be cross checked independently. For example there may be potential for combinations of low power tests with complementary failure modes. It is notable that enforcement of prohibitions requires tests which are more secure to false negative than false positive results, whereas ensuring quality of radiation processing might require the converse. The security with which labelling or commercial claims can be validated is sensitive to both modes of failure.

Quantitative tests become relevant in the context of a permitted but regulated process. The ability to verify stated or appropriate dose levels in irradiated products could also be an important quality assurance step for the clients of commercial irradiation plants. In this case it should be noted that the Codex dose specifications are for <u>average</u> dose level within a product batch, allowing for spatial variations of +- 50% within the batch. Tests which respond quantitatively over the 0.05 - 10 kGy dose range within this precision limit would be needed. Phenomena which require radiation calibration for individual samples may be useful here, as long as the calibration process can be applied retrospectively to the original radiation treatment.

Additionally the proscription of re-irradiation, except under special conditions (eg. incorporation of irradiated spices into compound foods for later treatment) suggests a need for tests which are sensitive to the time between successive exposures.

The Codex Alimentarius committee were originally pessimistic about the prospects of developing tests

> "Despite the many investigations designed to detect physical, chemical and biological changes in foods subject to ionising energy no satisfactory method for identifying food as having been irradiated has so far been developed".[4]

Similarly the UK ACINF committee in 1986 concluded

> "There are as yet no generally applicable chemical or physical tests which would be adequate for the enforcement of legal and commercial requirements and for investigation purposes, for determining whether a food or a food ingredient has been irradiated, and if so, at what dose".[6]

However since these comments were written there has been a renewal of interest in research on detection techniques. A number of physical, chemical and biological changes have been, or are being, investigated. To be successful research must be backed by demonstrably radiation induced effects which can be readily measured; the durability of signals over product shelf lifes should be investigated under a range of storage conditions, and the reliability should be assessed experimentally using representative sets of real food samples irradiated to food processing doses.

All successful identification techniques depend on three stage processes. The first stage, which takes place during, or immediately after, irradiation induces the effect. Thereafter a second storage stage is entered during which the phenomenon is preserved as a memory of the earlier ionising event. The third and final stage is the measurement. From the physical point of view these stages can be seen as energy transfer, energy storage and energy sensing/release, and the problem reduces to one of identifying and evaluating suitable associated metrological schemes.

4 LUMINESCENCE APPROACHES

Luminescence is the emission of light resulting from the relaxation of charge carriers in excited systems which have acquired excess energy by one means or another. Its association with ionising radiation is very long standing; indeed it was in the course of investigations of luminescence, using photographic recording techniques, that Henri Becquerel first discovered radioactivity in 1896. The notion that radiation makes things glow in the dark is not only firmly engraved in the popular imagination; it is also well founded from a phenomenological point of view.

There are two main classes of luminescence phenomenon; prompt and delayed forms. Delayed luminescence inevitably involves energy storage, where excited charge carriers are held in metastable states with excess energy compared with the unexcited system. In solid dielectric media there is a wealth of experience of such phenomena, associated with trapped electrons or ions localised at point defects in the host matrices. Charged molecular fragments can also be stored either by trapping in a lattice, or in principle by formation of metastable compounds. The first two stages of the ternary scheme outlined above are thus met by systems which store excited charge carriers by trapping processes. The final stage of read-out in delayed luminescence approaches involves the release of trapped charge from the storage states followed by transport and de-excitation with accompanying luminescence emission. Stimulated release processes, such as thermo-luminescence or photo-stimulated luminescence involve transfer of small amounts of energy to the trapped charge carrier population followed by release of larger quanta of energy in the luminescence relaxation.

Although there are many alternative physical techniques for measuring trapped charge - eg thermally stimulated exoelectron emission, thermally stimulated conductivity, electron spin resonance, optical density, - luminescence approaches have two main benefits : sensitivity and specificity. The sensitivity with which low levels of light emission can be detected, using the single photon counting technique is unparalleled by any alternative low cost instrumental approach - for example the relaxation of a few hundred charge carriers can be detected directly. This could not be achieved with optical density techniques, electrical methods or by microwave absorption (eg as in ESR). As far as specificity is concerned the recovery of optical quanta during relaxation stages itself can be used to confirm the energy storage resulting from radiation exposure. Again this direct denotation of the preceding ionising event is not matched in schemes which depend on detection of absorption bands or unpaired electrons.

Early attempts to utilise luminescence in the field include investigations of lyoluminescence[15,16] of saccharides and carbohydrates where chemically stored energy was released during solvation. Efforts to enhance the response of this rather delicate process using luminol chemical amplifiers[17,18,19,20,21] in application to herbs and spices appeared promising initially. However recent indications are that chemiluminescence (CL) signals obtained in this way are associated with the unstable reactions of free radicals in the period immediately following irradiation. It seems that reliable chemical storage processes with selective CL readout steps have not yet been identified in real foods. In addition to the stability problem there is also evidence that strong CL signals are associated with thermally induced free radicals. Therefore unless the possibility of heat treatment can be eliminated, using other indicators, even positive CL signals must be viewed with caution.

More promising approaches are thermoluminescence, where heat is used to stimulate the release of trapped charge, and photo-stimulated luminescence where tuned light sources are used. The first of these (TL) is a well established technique for radiation dosimetry using doped synthetic inorganic phosphors - the majority of routine personal dosimetry readings in hospitals or nuclear laboratories use TL of phosphors such as LiF or $CaSO_4$ for routine measurements in the μGy to mGy level. TL of naturally occurring minerals is also used for archaeological or geological dating. In these cases the integrated radiation dose due to long term exposure to low-level background is quantified using TL measurements of carefully separated pure mineral species. The time elapsed for such accumulation can then be determined using independent measurements of background radiation dose rates. Common natural mineral systems show a dynamic range of dose response from mGy to kGy.[22,23]

Thermoluminescence was first applied to the problem of identifying irradiated foods by the Munich ISR group,[24,25] originally stemming from attempts to enhance the sensitivity of CL signals from irradiated herbs and spices. Strong positive signals were reported when whole samples of irradiated herbs and spices were simply heated in a commercially available Harshaw TL dosimeter reader. Work in Munich at this time was focused on developing simple procedures whereby the intensities of TL from individual herbs and spices heated in this manner were compared with reference data sets for each product, measured before and after irradiation. An empirical approach was adopted based on selection of 10 different sources for each product, and used to define thresholds above which the level of TL was considered to be indicative of irradiation.

Independent investigations of irradiated potatoes and a few selected herbs and spices were undertaken in the TL laboratories of the Scottish Universities Research and Reactor Centre (SURRC) in 1986.[26] By removing small amounts of adhering minerals from the outside of potatoes the results of sprouting inhibition doses could be clearly detected, as shown in Figure 1.

Preliminary measurements of 4 herbs and spices at this time confirmed the presence of positive, although poorly reproduced, signals from samples irradiated to 10 kGy doses. The TL measurements involved heating powdered organic, and sometimes leafy samples to 450 or 500°C at ramp rates of 5°C s^{-1} in a dry nitrogen atmosphere. Attempting to heat organic samples in this manner is inherently irreproducible (because of thermal contact problems), and also rather messy. It was suspected that the signal might in fact be coming from a minor inorganic phase in the sample. However in view of the results from Munich and the preliminary runs at SURRC, a decision was taken to undertake detailed studies of the TL signals from whole samples. The objectives of this study were to establish the practical limitations of simple whole sample approaches to qualitative identification of irradiated herbs and spices, and also to investigate the origins of the signal bearing phases.

Measured October 1986

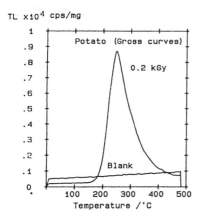
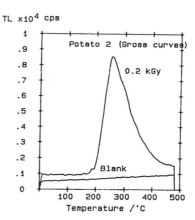

Figure 1 TL glow curves from soil washed from irradiated and unirradiated potatoes.

5 THERMOLUMINESCENCE OF WHOLE SAMPLES

An initial survey of some 67 different types of herbs, spices and seasonings was conducted using samples procured at local retail outlets in Scotland. The more common varieties were all represented by several separate samples purchased from different sources, and where possible originating from different countries. A total of 161 samples were studied in this first survey; a number which has subsequently been augmented considerably.

Samples were split into aliquots half of which were irradiated to a 10 kGy dose using a 200 TBq ^{60}Co source at SURRC, the other half retained as unirradiated controls. Dosimetry was controlled using R4034 perspex dosimeters cross calibrated to national standards. Packs of each sample and its associated control were stored in parallel at -20, 5, 30 and 55°C to provide series of aged samples under different storage conditions. Initial readings took place after 3 days of storage at 30°C.

Each sample and its associated control were prepared for TL measurement as follows. Coarse samples were gently powdered in a pestle and mortar, and leafy samples were finely chopped with a herb knife. Thereafter thin layers of each sample were deposited onto weighed stainless steel discs (0.25 mm thick, 1 cm diameter) which had previously been cleaned and coated with silicone grease spray. Excess material was shaken off, and the discs reweighed to determine sample mass. Roughly 5-15 mg of material adhered. All sample handling from the irradiation step onwards was conducted in subdued red lighting, in accordance with good practice in TL dating or dosimetry work.

TL measurements were taken with a SURRC research reader incorporating computer controlled temperature ramp and single photon counting. Each sample was heated twice at either 3° C s^{-1} or 6° C s^{-1} from room temperature to 500° C, the first glow to record TL, the second to record the black body background which was automatically subtracted. The photomultiplier (Thorn-EMI 9883QB) spectral bandpass was limited by colour filters to 350-450 nm.

The findings of this survey which have been discussed in further detail elsewhere,[27,28,29] revealed the following features. The majority of samples showed detectable thermoluminescence when irradiated. Unirradiated blanks showed low level background signals biased to higher temperatures than the radiation induced response. With the exception of those seasoning samples containing salt, all samples produced a broad TL glow curve, resulting from a continuum of trapping centres, and reminiscent of the glow shape familiar to those involved in TL dating as arising from feldspars and clay minerals. Typical examples of this form are shown in Figure 2. Those seasoning mixes containing common salt show a strong and characteristic signal from this component as shown in Figure 3.

The integrated TL signal strength, illustrated in Figure 4 using the peak signal integral from 200-210° C, shows marked variation, both in the levels observed from irradiated samples and unirradiated blanks. Radiation induced signals from all 161 samples span 4-5 orders of magnitude, while blank levels occupy 2-3. Taking signal strength alone as an indicator of whether samples have been irradiated would lead to successful identification of roughly 90% of positive signals from these freshly irradiated samples. However a significant minority of samples gave little or no TL on irradiation and would thus be classified as false negatives using this approach. There are also a few outlying samples from the unirradiated blank distribution, which have either been irradiated prior to purchase or are recorded as false positive readings in this simple approach.

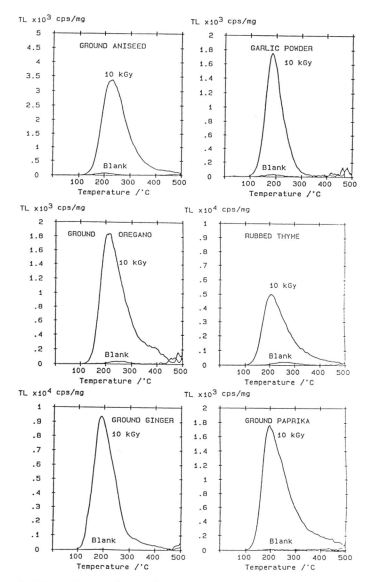

Figure 2 Examples of typical TL glow curves from whole samples of irradiated herbs and spices

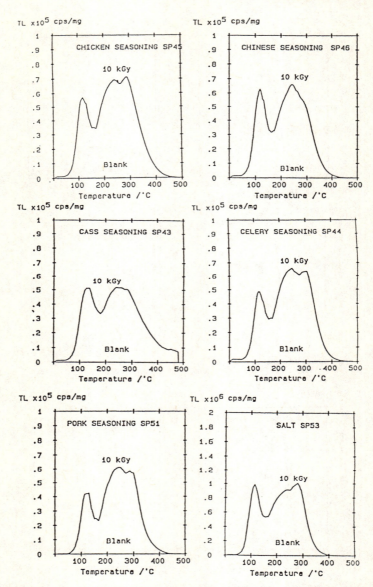

Figure 3 The characteristic glow curve from salt containing samples.

Further analysis of this data set has indicated[27] that there are partial correlations between some high signals and high blank levels, and that sorting samples out by morphological form can lead to a slight improvement in discriminating power using signal strength as a univariate classification parameter. For example spices derived from rhizomes, capsicums, and barks tend to have higher signal levels than peeled seeds or bulbs. Groupings of this type can improve whole sample discrimination to about 94%, but there are still problems with false negative results from low sensitivity samples, and with a high level of biological variability (up to one order of magnitude) for separate sub-samples of the same product.

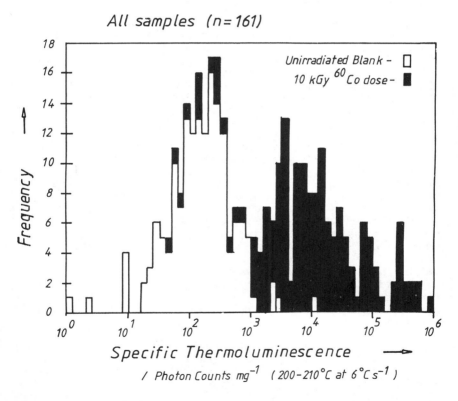

Figure 4 Sensitivity histogram showing the TL signal strength from 161 herbs spices and seasonings before and after irradiation to 10 kGy.

The Munich group have proposed a scheme based on the supposition that individual herbs and spices have their own TL characteristics. The average TL signal from 3 or 5 sub-samples, to reduce internal biological variability, is compared with an empirical threshold signal level defined from a reference set of samples for each variety.

As will be seen there are a number of problems with this approach when the origins and stability of the signal are fully considered. The SURRC data set does not support the view that individual varieties of herbs and spices have characteristic TL sensitivities. However the Munich approach has been applied in two interlaboratory trials with encouraging results, which at the very least show that TL results can be reproduced by different operators.[30,31] In the second of these 45 blind samples representing 9 different spices in sets of five were circulated to 11 different laboratories in sets of five of which it was known that at least one sample out of five was irradiated to 10 kGy, and that there was at least one unirradiated sample. The laboratories following the Munich protocol were able to identify all but one of the irradiated samples with complete reliability, a single example of dried mushrooms proving difficult in some cases.

The SURRC laboratory was successful in correctly classifying all 45 samples as irradiated or unirradiated using an approach based on glow curve shape and strength, and a comparison between aliquots measured as received, and after addition of a known 10 kGy laboratory dose. It was however noted the success rate was higher for these samples than expected on the basis of our reference data set for randomly purchased samples. It is possible that the sample selection may have been biased towards high sensitivity sample types. The study therefore can be taken to show that whole sample TL tests can be reliably reproduced from laboratory to laboratory providing that the samples themselves are amenable to analysis. This approach has been adopted for official food testing in the Federal Republic of Germany.

The view from SURRC is that whole sample measurements can be prone to false negative results, ie the failure to positively identify all irradiated samples. The performance - at probably better than 90-95% of random purchased samples is much better than many earlier approaches, but not at the level of confidence needed for legal or commercial requirements. Improvement of this performance level can be achieved when the origins of the TL signals are taken into account.

6 THE USE OF ENRICHED MINERAL SEPARATES

Clues as to the origins of the TL signals from whole samples of herbs and spices come from the glow curve shapes, the high variability in sensitivity, the lack of variety specific signatures, and the known TL sensitivity of irradiated silicates derived from soils. The TL response of common feldspars and clay minerals is sufficient for a few micrograms irradiated at kGy levels to contribute all the signals observed from whole samples.

A series of simple experiments was used to confirm the origins of the signal in adhering dust and dirt. Small samples of a few typical herbs and spices were immersed in an aqueous solution of sodium polytungstate mixed up to a density of 1.6 g cm^{-3}, and agitated in an ultrasonic bath. Subsequent centrifugation formed clearly separated layers, the organic matter of the bulk herb or spice floating while a small, and in some cases nearly invisible, quantity of inorganic material sank to the bottom of the tube. By decanting off the top layer the two parts were separated, washed and used to form new TL samples. Figure 5 illustrates the effect of this separation on the TL signals from the two phases, indicating clearly that the minute amount of inorganic material adhering to the surface is responsible for the vast majority of the TL signal. Further investigation of oleoresins and other organic compounds has not yielded any indications which lead us to suggest that there are other significant TL components in herbs and spices than those associated with the inorganic phase.

Before examining the improved performance which can be achieved using separated minerals, a number of corollaries of this finding to the whole sample measurement approach were explored. Naturally the features of high variability and lack of clear variety dependence are readily explained by the inorganic origin. Attempts were made to relate the whole sample results more clearly to the quantity of sample actually producing a signal contribution. Total and acid insoluble ash contents were measured for each of the samples involved in the original TL survey, and whole sample results re-normalised in terms of TL per unit mass relative to ash contents. This resulted in an improvement to the discriminating power of whole sample tests- the overlap

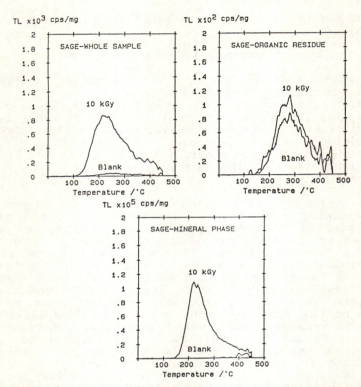

Figure 5 A comparison between whole sample, organic residue and inorganic TL fractions.

between blank and irradiated samples being significantly reduced. However the improvement was not sufficient to completely resolve all samples from all blanks. There are two reasons for this. The first is that the TL sensitivity per unit mass of silicates depends critically on the minor intrinsic and interstitial point defects within each mineral. Even pure alkali feldspar samples show a 2-3 order of magnitude variation in specific TL due to differences of geological and thermal history.[32] Therefore the weight normalised response per unit dose from irradiated samples shows a considerable spread. The second contributory factor is the so called "spurious" contribution to background readings resulting from the effects of heat on the porous organic matter of the spices. Despite conducting measurements in a pure nitrogen atmosphere the nature of the samples and their residual oxygen and moisture contents are partly responsible for a variable blank level.

Luminescence Detection of Irradiated Foods 43

The spurious effects due to organic samples can be eliminated by using separated mineral samples for TL measurements. In some cases wet sieving in water is sufficient to dislodge and concentrate the mineral fraction. The procedure which has been found more reliable at SURRC however is based on heavy liquid centrifugation as described above.

Simple TL signal strength tests based on separated minerals offer yet further improvements in signal to background ratio compared with ash normalised whole sample measurements. Nevertheless it was not until TL sensitivity was taken into account that it was possible to claim unambiguous discrimination between all irradiated and unirradiated samples. This was achieved effectively by re-calibrating each TL sample disc using a re-irradiation method.

7 THE USE OF RE-IRRADIATION

Failure to record a strong positive response can either be interpreted in terms of a lack of sensitivity to radiation, or a negative (ie unirradiated) result in a blind test. This can be resolved simply by exposing the sample to a known radiation source to measure the sensitivity. Genuine unirradiated samples show low signals but high sensitivity; whereas failure to respond to a calibrating dose indicates that the test is unable to reach a conclusion. It is only by completing the logical cycle of interpretation of tests for radiation in this manner that a reliable forensic test can be constructed.

With TL radiation calibration can be achieved in two ways. Either split aliquots are measured, one of which has been exposed to known dose, the other measured as received; alternatively if the first measurement does not destroy the sample then post-calibration can be applied.

Whole samples are destroyed by heating and therefore cannot be re-irradiated after measurement. Split aliquot techniques with whole samples have been investigated at SURRC - using the 45 blind samples correctly identified for the second ISH intercomparison.[29,33] Although this step is helpful in the context of whole sample measurements it is still limited by sub-sampling variability and spurious blank levels.

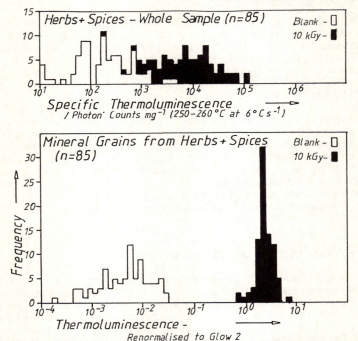

Figure 6 The effect of mineral separation and post-calibration on the power of TL to distinguish between 85 irradiated and unirradiated herbs and spices. Results from separated mineral grains have been re-normalised to the second glow response to a fixed 2.5 kGy gamma dose.

The most successful procedure which has so far been devised therefore comprises the combination of measurement of separated mineral samples which are individually re-irradiated and measured after the initial TL signals have been read. Figure 6 shows the impact which this approach has on the success with which 85 herbs, spices and seasonings can be correctly identified as irradiated or unirradiated compared with whole sample measurements. The mean TL signal from irradiated samples is 352 times that of the unirradiated controls; even the lowest positive signal is 25 times higher than the highest blank level (which was from a different sample).[35]

The benefits of the combined use of minerals and re-irradiation are in dramatically increasing signal to background ratios. The necessity of such an improvement becomes all the more apparent when signal stability is considered.

8 SIGNAL STABILITY

The stability of TL from silicate minerals has been widely studied in the context of TL dating studies.[32] The TL signals are composed of a pseudo-trap depth spectrum – higher temperature components generally originating from deeper traps within the crystal band gaps. Thermal fading from single trapping species may follow first order kinetics characterised by a mean life τ given by

$$\tau = s^{-1} \exp(E/kT_s)$$

where E is the trap depth, k is Boltzmann's constant, T_s is the storage temperature, and s a frequency factor.

For poly-mineral samples, or feldspathic minerals with broad distributions of trap depths losses due to thermal fading can in principle be described by summations of such first order components – therefore showing a composite non-exponential decay with time and complex storage temperature dependence. Fractional glow analysis coupled to Arrhenius transformation to analyse trap depths in these circumstances yield thermal activation energy spectra crossing from about 1 eV to 1.6 eV corresponding to the glow curve from 100°C to 500°C. Associated frequency factors are of the order of $10^{12} - 10^{13}$ s^{-1} resulting in mean life estimates at ambient temperature ranging from a few hours (for the 100°C ordinate) to over 10^7 years at the top of the glow curve. Thus the glow curve itself denotes a stability spectrum based on first order kinetics.

In addition to thermal losses however, a number of additional processes, including quantum mechanical tunnelling between trap and re-combination centre have been postulated,[35,36,37,38] to account for short term athermal losses originally called "anomalous" fading.[39] These proximity effects – if present – are expected to be more prominent at high doses than low doses for statistical reasons, and would give rise to a hyperbolic time dependent signal loss which does not depend markedly on sample storage temperature.

Given the certainty of multiple component signals, and the possibility of both thermal and athermal fading processes operating at low rates, empirical investigation of signal stability is essential. Long term stability tests have been conducted, initially using whole samples, but more recently with pure mineral examples. The procedure developed involved preparing parallel batches of irradiated and unirradiated controls stored at 4 different temperatures $-20, 5, 30,$ and $55°C$, and taking matched readings for samples and controls at logarithmically spaced time interval starting from 3.5 days after irradiation and proceeding so far for 2 years. An additional set of samples was spread in a thin layer and exposed to daylight at ambient temperatures to investigate the influence of optical bleaching.

The whole sample fading tests showed a certain amount of scatter, for reasons which have been indicated above. Nevertheless they have provided evidence that thermal losses dominate, although athermal processes may occur as a minor feature in frozen storage. The extent of signal loss with time depends both on glow curve temperature, and storage temperature - in keeping with expectations. Long term progressive signal losses were most clearly seen in the $200°C$ glow ordinate from samples stored at $55°C$ - the rate of loss at the $250°C$ ordinate being an order of magnitude lower. Losses at this latter glow temperature were of negligible importance at lower storage temperatures, and for samples with good initial signal to background ratios fading would not pose obstacles for correct identification. Figure 7 shows a whole sample fading test for Cayenne Pepper to illustrate most of these points.

Initial signal losses from samples exposed to light were more pronounced - and could erode confident judgements based on whole sample tests. However the rate of bleaching diminished rapidly and still left a durable unbleached component which may well prove to be sufficient to secure identification of samples exposed to high levels of illumination. For herbs and spices this problem does not arise since they are generally opaque and are clearly damaged beyond use or value by exposure to the levels of light needed to effect TL signals. However the influence of optical bleaching will need further consideration with respect to other sample types which may be investigated.

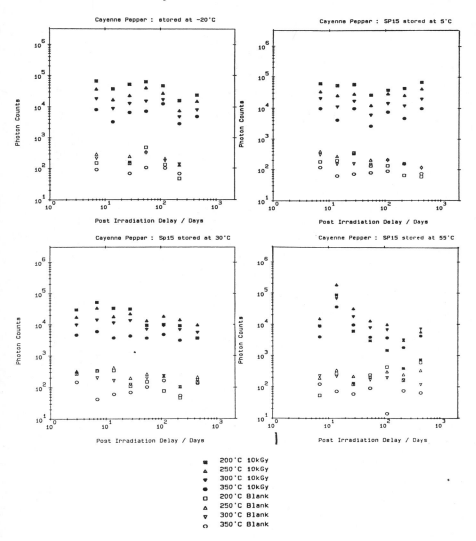

Figure 7 Long term fading results from Cayenne Pepper. The symbols show the TL signal at different glow curve temperatures remaining in whole samples. Each graph shows the results of parallel tests at different storage temperatures.

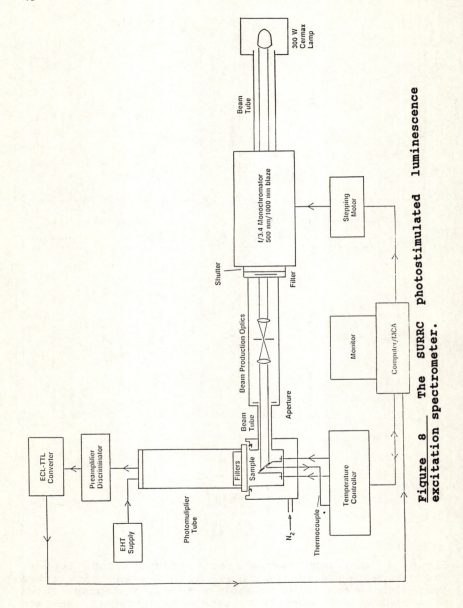

Figure 8 The SURRC photostimulated luminescence excitation spectrometer.

9 RANGE OF APPLICABILITY

The mineral debris responsible for TL in herbs and spices occurs ubiquitously on all foodstuffs which have been exposed to wind and soil. In principle therefore all fruits and vegetables - in addition to the 52 herbs and spices already investigated, may also prove to be reliably labelled with luminescence dosimeters. As well as the potatoes referred to originally, investigations at SURRC have shown that even avocado pears retain sufficient mineral material to be amenable to TL testing. Strawberries have also been investigated briefly in Germany [40] - once again the origin of the signal being most likely to come from inorganic matter adhering to the skin. When extending sample range towards a wider range of agricultural produce it is evident that more attention to the effects of optical bleaching will be needed, and this is an important area for further research. There are also a number of important bioinorganic materials - including bones, shells and chitin which show positive TL when irradiated. However their porosity and intimate interlacing of organic components produce limitations with spurious luminescence.

The strengths of TL testing based on separated mineral extracts coupled to calibration by re-irradiation are the high level of confidence which can be achieved with good laboratory quality control, and the wide potential range of application. These features combine to put TL in the forefront of currently available techniques for identification of irradiated foods, and suggest that it will be applied in commercial and forensic contexts for some time. However it should be acknowledged that the sample preparation steps are demanding, and that the need to use laboratory re-irradiation to confirm negative results may be a source of inconvenience to some analysts. Looking to the future it is desirable to develop techniques which can overcome these limitations, and which can also measure trapped charge populations in bioinorganic matter without the disruptive effects of sample heating.

10 PHOTOSTIMULATION

The possibility of using photostimulation to read the luminescence out as an alternative to heating samples is being explored with a view to eventually overcoming some of the limitations mentioned above. In this approach to the problem the energy required to release trapped charge prior to production of luminescence is supplied to the sample by illumination. A spectrometer (Figure 8) has been constructed at SURRC comprising a 300W Cermax Xe lamp coupled to a stepping motor driven f 3.4 monochromator and beam producing optics to illuminate samples with tunable photon beams. The sample chamber comprises a temperature controlled stage capable of operation from $-130°C$ to $500°C$ and viewed by a Thorn EMI 9883 QB single photon counting photomultiplier coupled through to a computer based multichannel scaling system. Optical filters are used to define the detection spectral window, and for order sorting through the monochromator. This system has been configured initially to investigate the excitation spectra from irradiated samples of various types.

When the detection window is spectrally limited to 300-350 nm using dense combinations of optical filters it is possible to excite **anti-stokes luminescence** from irradiated samples illuminated with longer wavelengths from 450-950 nm. This really means that the quantum energy released in the output photon is greater than the stimulating quantum energy - the difference being made up solely from the stored energy left behind after irradiation. The specificity with which this phenomenon denotes energy storage and irradiation is unmatched elsewhere. Figure 9 illustrates the anti-stokes luminescence from an irradiated pure feldspar - the UV output being excitable throughout the visible and near IR regions. Unirradiated samples show background signals which are solely due to photomultiplier dark count, or any filter leakage from the stimulation source.

The dosimetric response of these signals from 100 Gy to 10 kGy has been confirmed experimentally, and preliminary thermal annealing experiments have suggested that the IR peaks are associated with a similar stability range as the TL glow curve regions used to identify irradiated herbs and spices. This being so experiments were successfully conducted to identify the IR peaks firstly in pure mineral separates extracted from irradiated herbs and spices, and then directly from whole samples which had previously shown positive TL signals.

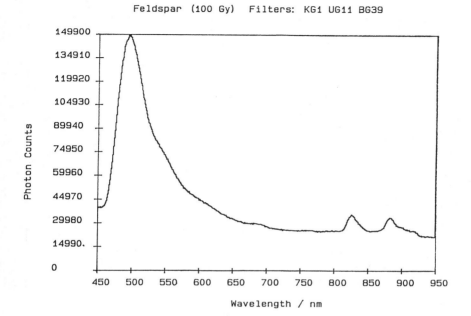

Figure 9 PSL excitation spectrum for microcline feldspar.

Since the readout by PSL spectroscopy does not destroy the sample it is possible to apply single-aliquot re-irradiation methods to calibrate unknown samples. Under this procedure an unirradiated sample will give a low signal on first measurement followed by a dramatic signal increase when re-measured after irradiation whereas an irradiated sample will only show a modest increase in signal level. Figure 10 illustrates this approach for herbs and spices - suggesting that it will be possible to use a PSL approach of this kind to achieve reliable detection of irradiation without mineral separation. These are very promising early results from this new approach and suggest that PSL will find its place amongst the repertoire of accepted techniques once a broader suite of samples have been investigated. Initial tests with bioinorganic samples have also yielded encouraging results.

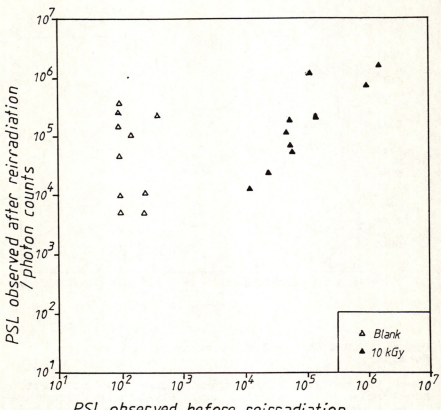

Figure 10 PSL measurements of unseparated herbs and spices using a re-irradiation calibration technique.

11 SUMMARY AND CONCLUSIONS

In summary the need for detection methods is clearly established at a time of growing international trade in irradiated products and continuing diversity of national regulation of the process. Of the many approaches which are under investigation in various laboratories, luminescence techniques are developed to a high level of reliability for many product types. Thermoluminescence measurements in particular can be used for unambiguous qualitative identification of irradiated herbs and spices, although sample preparation and re-irradiation are currently needed to achieve forensic levels of reliability.

There are prospects for further extension of the product ranges for TL tests to cover most fruits and vegetables, and potentially to achieve quantification for samples whose storage histories can be defined within limits. Developments based on photostimulation also hold prospects for rapid testing without the need for extensive sample preparation, and to extend the sample range through to shellfish, and meats containing bone fragments.

REFERENCES

1. P. Elias and A. J. Cohen, "Recent Advances in Food Irradiation", Elsevier Biomedical Press, Amsterdam, 1983.

2. E. S. Josephson and M. S. Peterson, "Preservation of Food by Ionising Radiation", (3 vols), CRC Press, Boca Raton, Florida, 1983.

3. W. Urbain, "Food Irradiation", Academic Press, London, 1986.

4. FAO/WHO, "Wholesomeness of Irradiated Food", WHO Technical Report 659, HMSO, London, 1981.

5. FAO/WHO, Codex Alimentarius Vol. XV, Ed.1, Codex Standard 106, WHO, 1983.

6. ACINF, "Report on the safety and wholesomeness of irradiated foods", HMSO, 1986.

7. ACINF, "Response to comments received on the Report on the Safety and Wholesomeness of Irradiated Foods", DHSS, London, 1987.

8. FDA, "Irradiation in the production, processing and handling of food, Final rule 21 CFR, part 179", Fed. Regist., $\underline{51}$(75), 13376, 1986.

9. CEC, "Report on the Wholesomeness of Foods Irradiated by suitable procedures", CEC, Brussels, ISBN 92-825-6983-7, 1987.

10. IAEA, *Food Irradiation Newsletter*, 1987, $\underline{11}$(1), 1.

11. WHO, "Food Irradiation: A technique for preserving and improving the safety of food", WHO, 1988.

12. Statutory Instruments, "The Food (Control of Irradiation) Regulations, 1967, "England and Wales: SI 1967/385, Scotland: SI 1967/388, Northern Ireland: NI 1967/51.

13. Statutory Instruments, "The Food (Control of Irradiation) (Amendment) Regulations, 1967", England and Wales: SI 1972/205, Scotland: SI 1972/307, Northern Ireland: NI 1972/68.

14. House of Lords, Select Committee on the European Communities, "Irradiation of Foodstuffs", HMSO, 1989.

15. K. V. Ettinger, J. R. Mallard, S. Srirath and A. Takavar, Phys. Med. Biol., 1977, 22, 481.

16. K. V. Ettinger, J. R. Mallard, S. Srirath and A. Takavar, "Food Preservation by Irradiation", IAEA, Vienna, 1978, Vol 2. p. 345.

17. W. Bögl and L. Heide, Fleishwirtschaft, 1984, 64, 1120.

18. W. Bögl and L. Heide, Radiat. Phys. Chem., 1985, 25, 173.

19. L. Heide and W. Bögl, Z. Lebensmitt. Untersuch. Forsch., 1985, 181, 283.

20. L. Heide and W. Bögl, Proc. 4th European Conf. Food Chemistry, 255, ISBN 82-90394-17-9, 1986.

21. L. Heide and W. Bögl, Int. J. Food Sci. Technol., 1987, 22, 93.

22. D. C. W. Sanderson, R. J. Clark, C. Slater and K. J. Cairns, "TL Dating using Alkali Feldspars: High Dose Characteristics and Stability estimates", in "Long and Short Lower Age Limits in Luminescence Dating", Research Laboratory for Archaeology, Oxford University, 1989.

23. D. C. W. Sanderson, P. A. Clark, A. B. Dougans and J. Q. Spencer, "TL Dating using Alkali Feldspars: Sensitivity Range and Minimum Detectable Dose", in "Long and Short Lower Age Limits in Luminescence Dating", Oxford. 1989.

24. L. Heide and W. Bögl, "Die Messung der Thermolumineszenz - Ein neues Verfahren zur Identifizierung strahlenbehandler Gewurze", Institute für Strahlenhygeine, Heft 58, 1984.

25. L. Heide and W. Bögl, Fresenius Zeitschrift Analytische Chem., 1985, 320, 283.

26. D. C. W. Sanderson and J. A. Izatt, "Luminescence Methods for Determining Applied Dose in Irradiated Foods", in "Prospective methods for identifying irradiated foods", Manchester, 1987.

27. D. C. W. Sanderson, C. Slater and K. J. Cairns, "Development of Luminescence Tests to Identify Irradiated Foods", Progress Report 1, N384, MAFF, 1988.

28. D. C. W. Sanderson, C. Slater and K. J. Cairns, "Development of Luminescence Tests to Identify Irradiated Foods", Progress Report 2, N384, MAFF, 1989.

29. D. C. W. Sanderson, C. Slater and K. J. Cairns, Radiat. Phys. Chem., 1989, 34(6), 915.

30. L. Heide, H. Delincee, D. Demmer, D. Eichenauer, H. U. v Grabowski, K. Pfeilsticker, H. Redl, M. Schilling and W. Bögl, "Ein erster Ringversuch zur Identifizeirung Strahlenbehandler Gewurze mit Hilfe von Lumineszenzmessungen", ISH Heft 101, 1986.

31. L. Heide, J. Ammon, J. Beczner, H. Delincee, D. Demmer, D. Eichenauer, H. U. v Grabowski, R. Guggenberger, M. Guldborg, W. Meier, K. Pfeilsticker, H. Redl, D. C. W. Sanderson, M. Schilling, A. Spiegelberg and K. W. Bögl, "Thermolumineszenz und Chemilumineszenz Mesungen zur identifizerung strahlenbehandelter Gewurze", ISH Heft 130, 1989.

32. D. C. W. Sanderson, Nuclear Tracks, 1988, 14(1/2), 155.

33. D. C. W. Sanderson, C. Slater and K. J. Cairns, "Thermoluminescence Measurements of samples from the Second ISH Ringversuch:, SURRC Report, 1988.

34. D. C. W. Sanderson, C. Slater and K. J. Cairns, Nature, 1989, 340, 23.

35. G. F. C. Garlick and I. Robinson, "The Thermoluminescence of Lunar Samples" in "The Moon", ed. S.K. Runcorn and H.C. Urey, I.A.U., 1972.

36. R. Visocekas, T. Ceva., C. Marti, F. Lefaucheux and M. C. Robert, Physics Status Solidi, 1976, A35, 315.

37. R. Visocekas, M. Ouchene and B. Gallois, Nucl. Instrum. Meth., 1983, 214, 553.

38. R. H. Templer, "A New Model for Anomalous Fading" Ch.6, D. Phil Thesis, Oxford University, 1986.

39. A. G. Wintle, Nature, 1975, 245, 143.

40. K. W. Bögl, Bundesgesundhbl, 1989, 9/89, 388.

Changes in DNA as a Possible Means of Detecting Irradiated Food

D. J. Deeble, A. W. Jabir, B. J. Parsons, C. J. Smith and P. Wheatley

NORTH EAST WALES INSTITUTE, CONNAH'S QUAY, CLWYD CH5 4BR, UK

1 INTRODUCTION

Food is a complex system consisting of proteins, carbohydrates, fats, salts, vitamins and water. Such a definition would probably represent an immediate response as to the chemical content of food. However, one important class of molecules, the nucleic acids, is not included in this list. Since most foods are derived from living organisms and all living organisms contain nucleic acids, food contains nucleic acids. Indeed the biological 'raison d'etre' of foods such as fruits, cereal grains and vegetables is to act as containers for the plant's DNA (deoxyribonucleic acid) and to ensure that this DNA is transported to a site where a new plant can grow. Although RNA (ribonucleic acid) will also be present in many foods, the amount will be smaller than that of DNA and its lower molecular weight and higher susceptibility to degradation make it less convenient for study than DNA. Consequently, only DNA will be considered although many radiation products are common to both types of nucleic acid.

Ionising radiation is particularly efficient in its ability to kill life forms, for example the lethal radiation dose (ca. 4Gy) estimated for humans, is, in energy terms, equivalent to raising the victim's body temperature by less than one thousandth of a degree centigrade, less than the increase caused by drinking a hot cup of tea. This property of ionising radiation, to destroy life forms with a relatively low energy input and hence a relatively small amount of chemical change, is, of course, a major reason for its use in food processing. Thus the numbers of micro-organisms and insects in foods can be drastically reduced without any appreciable alteration to the taste or texture of the food. As well as reproductive cell death, irradiation (throughout this paper only ionising radiation is considered) causes mutagenesis and transformation (cancer), all of these effects are associated with DNA damage. It is now widely accepted that DNA is the major cellular target for irradiation, in that it is radiation induced DNA damage which is responsible for the acute effects observed. In view of its prime importance in radiobiology, it is hardly surprising that the radiation chemistry of DNA has received considerable attention (some reviews are given in refs.[1-5]).

The biological consequences of radiation induced damage to a cell's DNA are, as already mentioned, catastrophic and it has been, and continues to be, the task of the chemist to identify the chemical nature of this damage and the mechanisms involved in its formation. The aim of this paper is to examine whether radiation damage to a food's DNA can be utilised as a marker to show that the food has been irradiated. Before this is possible, it is necessary to examine not only the types of damage produced but also how these might be affected by the conditions of irradiation. Therefore, some of the products and mechanistic aspects of the radiation chemistry of DNA will now be presented. This will be followed by a presentation of some of the approaches being adopted in an attempt to apply these findings to the development of suitable assays for irradiated food.

2 RADIATION INDUCED DNA DAMAGE

The Structure of DNA

DNA is a double stranded polymer, the strands being intertwined to give the well-known double-helix.[6] Each strand is made up of a deoxyribose phosphate backbone with the C(1') position of each deoxyribose being substituted by a nucleobase (Figure 1). There are four common DNA bases, the pyrimidines, thymine and cytosine and the purines, adenine and guanine. The C(1') base substituted deoxyribose moieties are called deoxyribonucleosides and the corresponding orthophosphate esters (5' or 3') are called deoxyribonucleotides. In DNA, the two polydeoxyribo-nucleotide chains are held together by hydrogen bonding between specific base pairs, adenine and thymine form one pair, while guanine and cytosine form the other.[6]

The Indirect Effect

Ionising radiation, as its name implies, interacts with the electrons of matter through which it passes, so that some of its energy is transferred to these electrons and ionisation occurs. The

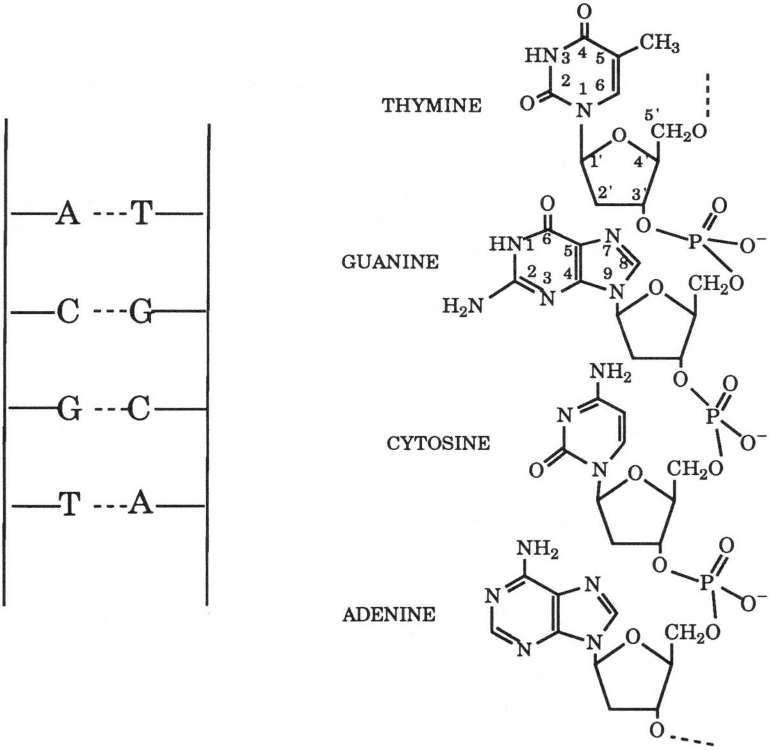

Figure 1. Structure of DNA. On the left is a diagrammatic representation of double stranded DNA. On the right is the structure of a section of one of the strands.

ejected electrons usually possess sufficient energy to cause the ionisation and excitation of nearby molecules. The ejection of an electron from a molecule results in the formation of a cation having an unpaired electron. These cations are unstable and undergo further reactions, the products of which also often contain an unpaired electron. Molecules possessing an unpaired electron are known as free radicals and are usually highly reactive.

On irradiating a particular system, a given 'target' can be chemically altered either by the direct ionisation of molecules of the 'target', the 'direct effect' (see below) or by the attack of reactive species (e.g. free-radicals) formed by the ionisation and excitation of the molecules surrounding the target, the 'indirect effect'. The probability of a high energy photon interacting with a particular atom depends on its electron density, in biological systems the most common atoms present are C, O, N, and H. These atoms are distributed more or less evenly so that initial ionisation will occur more or less randomly throughout the system. Biological systems contain a high percentage of water (typically 60%), and a comparable percentage of the initial energy deposition (i.e. ionisation) must be in the water. The radiation chemistry of dilute aqueous systems is now well understood (cf. [1-7]). Here, almost all of the energy (>99%) is absorbed by the water with the formation of water derived radicals and products, (reaction (1)) and these may then react with any solutes present. By varying the irradiation conditions (e.g. the addition of specific radical scavengers) it is possible to obtain essentially uniradical systems, so that the reactions of a chosen radical e.g. the OH radical (see reactions (1) and (2)) or the solvated electron (e_{aq}^-) (reactions (1) and (3), R = alkyl group) can be studied. The alcohol radical produced in reaction (3) is much less reactive than e_{aq}^- and should not interfere. In N_2O - saturated solutions 90% of the initial radicals are OH, the remaining 10% being H:

$$H_2O \xrightarrow{\text{radiation}} {}^\bullet OH + H^\bullet + e_{aq}^- + H^+ + H_2 + H_2O_2 \quad (1)$$

$$N_2O + e_{aq}^- \xrightarrow{H_2O} {}^\bullet OH + OH^- + N_2 \quad (2)$$

$$RCH_2OH + {}^\bullet OH \longrightarrow RCH^\bullet OH + H_2O \quad (3)$$

DNA is heterogeneous and the water derived radicals can react with any of the component nucleotides. The complexity of the system has been reduced by investigating these components separately. The pyrimidines have in the past, received the most attention e.g. refs. [1-4, 8-10] although investigations have also been performed on the purines [11-13] where considerable progress in understanding the processes occurring has recently been made.[14] It is not the purpose of this paper to attempt to review the large amount of research performed on the free radical chemistry of the nucleobases and their derivatives.

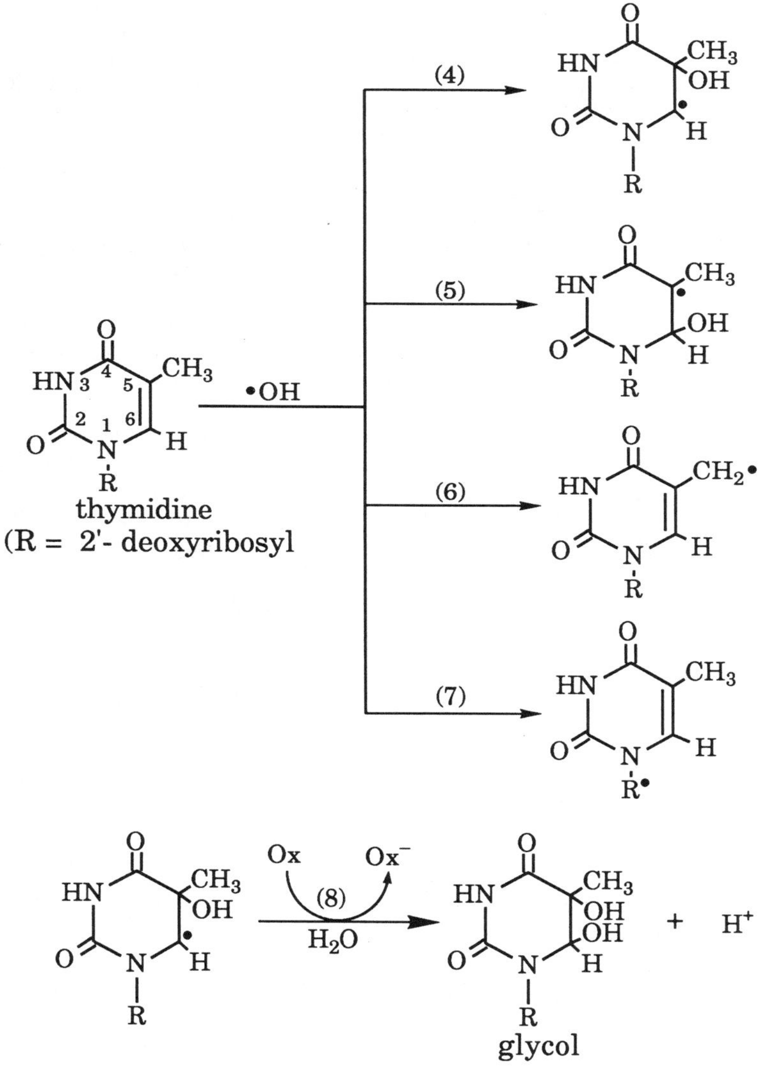

Scheme 1. The reaction of OH radicals with thymidine and the formation of thymidine glycol

Therefore, only thymidine will be considered since it is hoped to use two of the radiation products of thymidine as markers for radiation treatment (see below).

OH radicals react with thymidine mainly by addition at the C(5) - C(6) double bond, although some H abstraction occurs at the C(5) - methyl group this is believed to occur to an extent of <5% (see scheme 1, reactions (4), (5), (6), (7)).[1-15] Some H abstraction from the sugar moiety may also take place, this is also likely to represent a small fraction (<10%) of the overall radical yield cf. in the polynucleotide, poly(U), direct OH radical attack on the ribose accounts for ca. 7% of the total yield of ·OH.[16] In DNA, many of the deoxyribose radicals eventually lead to strand scission.[17] In poly(U), base radicals abstract H atoms from the ribose, thus effecting an efficient transfer of the radical site from the base to the sugar.[18,19] It is not yet known whether such processes occur in DNA.

The thymidine OH-adduct radicals formed in reactions (4) and (5) have different redox properties, the 5-OH, 6-yl radical being more easily oxidised than the 6-OH, 5-yl radical.[15] On oxidation of the 5-OH, 6-yl thymidine radical, either by reaction with a more oxidising radical or with an oxidant such as a transition metal ion e.g. Cu^{2+}, which is present in the cell nucleus, thymidine glycol will result (Scheme 1, reaction(8)). A similar reaction could be envisaged for the 6-OH, 5-yl radical, despite it being less oxidisable.

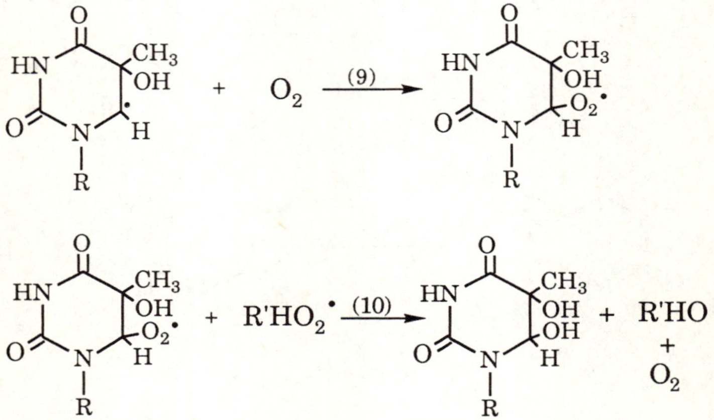

Scheme 2. An example of the formation of thymidine peroxyl radicals and their conversion to thymidine glycol.

In the presence of oxygen all of the thymidine radicals are rapidly converted into their corresponding peroxyl radicals as illustrated in Scheme 2, reaction (9), for the 5-OH, 6-yl radical. These peroxyl radicals can react with other peroxyl radicals (or with each other), in the case of the thymidine-6-peroxyl radical, thymidine glycol would be among the products formed. (Scheme 2, reaction (10) cf. ref.[20]).

Thymidine glycol is thus produced both in the presence and absence of oxygen on exposing aqueous solutions of thymidine to radiation. Studies on DNA also found thymidine glycol to be formed in the presence and absence of oxygen.[21,22]

On reaction of the solvated electron with thymidine, the thymidine radical anion is formed which in the presence of a proton donor (phosphate buffer) can be converted into the 6-H-adduct radical (see Scheme (3)).[23,24] The same radical is produced by the addition of an H-atom at C(6), the 5-H-adduct being formed by H-atom addition at C-5. H atom addition to thymidine proceeds mainly (ca.60%) by addition at C(6).[23] On reduction or by abstracting an H atom (from another radical or a neighbouring molecule) these radicals are converted to dihydrothymidine (cf. Scheme 4, reaction (17)). In fact, dihydrothymidine is a major product in irradiated deoxygenated thymidine solutions.[25]

$$O_2 + e_{aq}^- \longrightarrow O_2^{\bullet -} \quad (18)$$

$$O_2 + H^\bullet \longrightarrow HO_2^\bullet \quad (19)$$

$$HO_2^\bullet \xrightleftharpoons{pK_a = 4.8} H^+ + O_2^- \quad (20)$$

The rate constants for the reaction of oxygen with both e_{aq}^- and H (reactions (18), (19), (20)) are very high (ca.10^{10} dm^3 mol^{-1} s^{-1}), in addition the H-adduct radicals of thymidine also react rapidly with oxygen (k ca. 10^9 dm^3 mol^{-1} s^{-1}) hence in fully aerated systems, the yields of dihydrothymidine will be severely reduced.

The 'Direct' Effect

Ionisation of deoxyribose and the phosphodiester linkages will generally give rise to a DNA strand break.

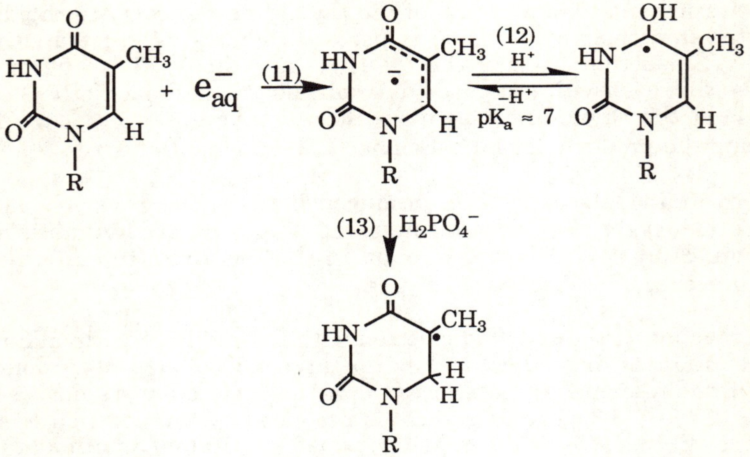

Scheme 3. The reaction of thymidine with the solvated electron to form the thymidine radical anion which can protonate rapidly at oxygen (reversible) and more slowly at carbon (irreversible).

When the bases are ionised a base radical cation and an electron will be produced. The 'electron sink' in DNA is thymine and e.s.r. studies on frozen DNA solutions have shown that thymidine radical anions are present, (along with guanine radical cations).[26-28] It appears that the thymidine radical anion in DNA protonates at C(6) to give the C(6)-H-adduct (this also occurs in aqueous solution in the presence of a proton donor, see above) and again this would produce dihydrothymidine on reduction or by H-atom transfer.

Direct ionisation of thymine in DNA produces the thymidine radical cation, recent studies have shown that deprotonation from N(3) can then occur (pK_a = 3.6, Scheme 5, reaction (21)) with the formation of an N-centred radical.[29] The radical cation decays by reaction with water (Scheme 5, reaction (22) and (23)). In DNA the N(3) of thymidine is H-bonded to the N(1) of adenosine and since the pK_a of adenosine is 3.5, complete deprotonation of the radical cation of thymidine seems unlikely, nucleophilic attack by water (reaction (22)) gives the 6-OH adduct radical which could lead to the eventual formation of thymidine glycol.

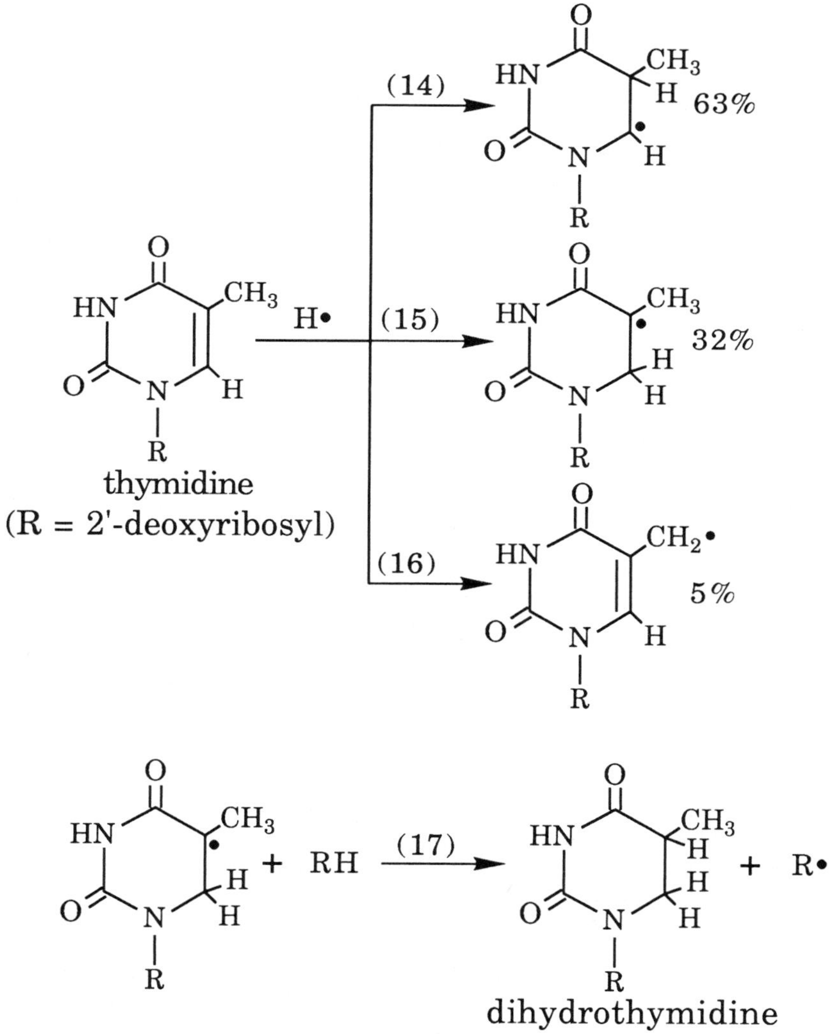

Scheme 4. Reactions of H atoms with thymidine and the conversion of the C(6)-H adduct to dihydrothymidine

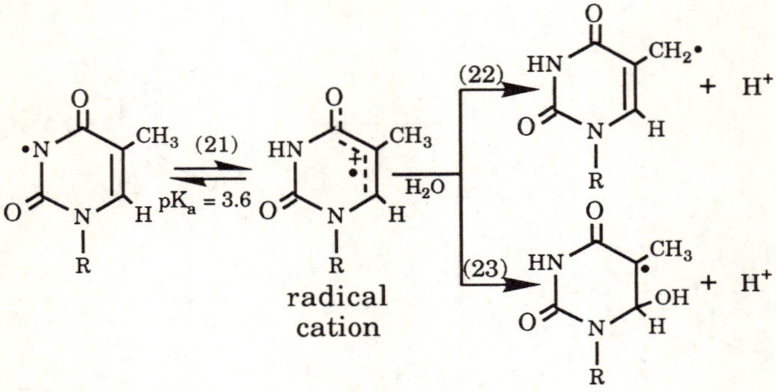

Scheme 5. Reactions of the thymidine radical cation.

Therefore, dihydrothymidine and thymidine glycol are produced by both the 'direct' and 'indirect' effects of radiation. Glycol is formed in both the absence and presence of oxygen, the yields of dihydrothymidine are likely to be reduced if oxygen is present during irradiation.

3 SOME METHODS OF DETECTING DNA DAMAGE AND THEIR PROSPECTIVE APPLICATION IN ASSAYS FOR IRRADIATED FOOD

In the previous section, specific mechanisms for the formation of two of the many radiation products of thymidine were presented. However it is clear that in DNA any of the bases can be modified by irradiation and that many sugar derived products will also be formed. As mentioned already, damage to the sugar frequently manifests itself as strand breakage. Consequently on exposure to radiation, DNA undergoes base modifications and sugar damage with the appearance of both single and double strand breaks.

A major problem in seeking a specific radiation product as a 'marker' for irradiation is that at the doses likely to be permitted in food processing (<10 kGy) the yield of any single product will be very low (<0.1 μmol kg^{-1}). Any successful assay will, therefore, have to be extremely sensitive. One means of overcoming the problem of sensitivity is to devise an assay which measures more than one single radiation product. Some potential assays will now be given in order of increasing specificity.

Overall Disruption of the DNA Double Helix

In native DNA, the bases are inside the double helix and are to a large extent protected against attack by reactive solute molecules. On irradiation, the double helix is disrupted by the appearance of single and double strand breaks and base damage, all of which cause partial unwinding of the helix and exposure of the bases to attack by solutes as illustrated in Figure 2.

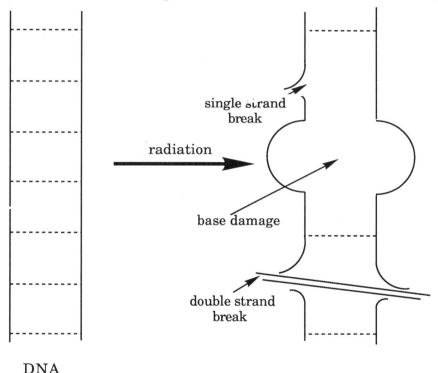

Figure 2. The gross effects of irradiation on the DNA double helix.

Formaldehyde reacts with the DNA bases by adding to the primary and secondary amino groups. In native DNA access to these sites is blocked and the reaction is slow. In solution DNA is a dynamic structure and 'breathes' i.e. a continuously changing section of the double helix opens up and recloses (for a review on DNA structure see ref.[30]). If a formaldehyde molecule is correctly situated it can react with a base during the time it is exposed, once the reaction has taken place, the helix cannot reclose at that point and bases on either side remain open to attack by formaldehyde. Thus from the initial site of formaldehyde addition the double helix is 'unzipped' in both directions.

Around the sites of damage in DNA, the bases are exposed to solute attack, therefore formaldehyde is able to react more rapidly with damaged DNA.[31,32] Figure 3 shows the spectra of native DNA, denatured DNA (i.e. the DNA strands have been separated by heating the solution, so that single stranded DNA only is present) and the DNA:formaldehyde complex. The reaction of DNA with formaldehyde can conveniently be followed by monitoring the

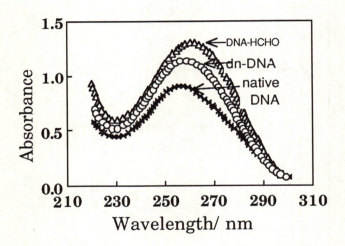

Figure 3.   UV - Spectra of native DNA, denatured DNA and the DNA-HCHO complex. The concentration of DNA (calf thymus) is the same for each spectrum.

absorbance increase at 270nm. In native DNA the bases are 'stacked' above each other and molecular orbital interactions between the bases are facilitated. The result of these interactions is that the extinction coefficient (in the u.v.) of native DNA is smaller than that of denatured DNA where the bases are now 'unstacked'. The increase in absorbance (e.g. at 260nm) on going from native to denatured DNA is known as the hyperchromic effect. From Figure 3 it can be seen that the extinction coefficients of thermally denatured DNA and the DNA-formaldehyde complex are almost identical at ca. 245nm. Thus the absorbance change at 245nm is due to the hyperchromic effect. Damaged DNA is partially denatured and the size of the hyperchromic change on complete denaturation will be smaller than for undamaged DNA. In contrast, the rate of reaction of formaldehyde with damaged DNA is faster than with native DNA (see Figure 4). A convenient measure of this rate is the absorbance increase 10 min after the reaction was started (DNA and formaldehyde were mixed at 4°C in an optical cuvette which was placed in a thermostatted cell block (10°C) inside a spectrophotometer, the reference contained DNA but no formaldehyde. Zero reaction time was taken as the point where water from a 60°C water bath was circulated through the cell block).

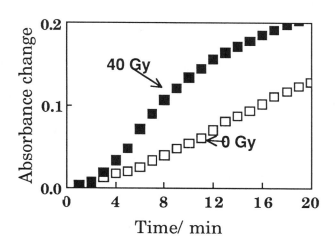

Figure 4. The change in absorbance at 270 nm as a function of time as formaldehyde reacts with either unirradiated or irradiated DNA (N_2O-saturated aqueous solution).

By taking the ratio (R) of the total absorbance change at 245 nm (i.e. the hyperchromic effect) divided by the absorbance change at 270 nm after 10 min reaction time, a quantity is obtained which is independent of the DNA concentration used and which is a sensitive measure of DNA damage.

Results on calf thymus DNA, irradiated in aqueous solution, and DNA extracted from irradiated wheat are given in Table 1. The method used for DNA isolation from wheat is relatively simple, details are given elsewhere.[33]

From Table 1, it can be seen that the value of R is strongly affected by radiation, as expected the DNA is considerably more radiosensitive in solution than in wheat (this is partly due to the higher concentration of the DNA used in the solutions compared to that present in wheat). Nevertheless, there is a measurable difference between unirradiated and irradiated wheat.

Table 1   Effect of irradiation on the ratio (R) of the overall hyperchromicity change (245 nm) to the absorbance change after 10 min of the formaldehyde reaction (270 nm).

	Calf Thymus DNA					Wheat DNA				
Dose/Gy	0	10	20	30	40	0	250	500	750	1000
R	2.46	1.66	1.12	1.07	0.91	2.34	2.28	2.14	2.05	1.86
R(irr.)/R(unirr.)	1.0	0.67	0.46	0.43	0.37	1.0	0.97	0.91	0.88	0.79

Although alterations in the value of R cannot be uniquely attributed to the effects of radiation in the particular case of dry cereals (or other dry seeds) where the DNA is stable for many years any detectable change in R would merit further investigation into the history of the sample.

DNA Chain Scission.

Both single and double strand breaks are produced in DNA by the action of radiation and a number of research groups have looked at the possibility of using these breaks as a means of detecting irradiated food.

Techniques such as membrane filtration, radioactive labelling of chain ends, and gel electrophoresis of mitochondria DNA have all proved successful in distinguishing various irradiated foods from their unirradiated controls.[34-36] One problem encountered is strand breakage caused by freezing and then thawing the food since this could obviously produce spurious results, although the absence of

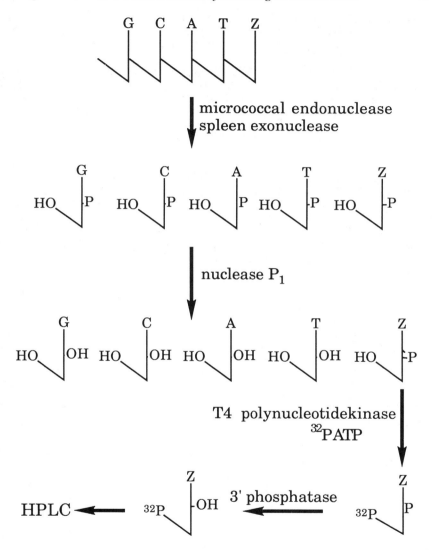

Scheme 6.   DNA post-labelling technique.

strand breaks would still signify non-exposure to irradiation. The use of mitochondrial DNA, which apparently remains intact during freeze/thaw cycles overcomes this difficulty.

Chain breaks are not only formed by irradiation e.g. numerous endonucleases exist, so again the discovery of excessive breaks would serve as a warning bell but probably could not be taken as definite proof of irradiation.

Non-Specific Base modifications.

Many different products are derived from the bases when DNA is irradiated. This is well illustrated by simply considering thymine alone, where on radiolysis in aqueous aerated solution 23 products have been identified.[37]

The DNA post-labelling technique provides a 'finger print' assay for any modified bases which might be present. In essence, the method involves extracting the DNA from the sample under

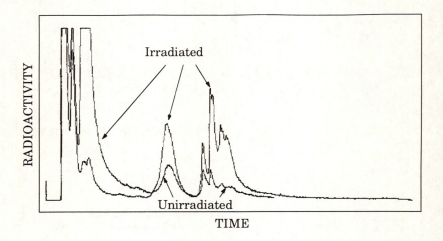

Figure 5.   HPLC from the DNA post-labelling technique applied to DNA isolated from unirradiated and irradiated (5 kGy) chicken.

investigation, enzymatically digesting the DNA to yield the 3 deoxyribonucleotides, specifically labelling base modified deoxyribonucleotides with ^{32}P and analysing using HPLC (or TLC) with an on-line radiochemical detector (see Scheme 6).

This technique has been successfully applied to detecting DNA damage in human mucosal cells induced by tobacco combustion products[38] and in the liver cells of fish living in polluted waters.[39] Recently the post-labelling assay has been applied to the problem of detecting irradiated food by Drs. Hoey and Swallow working at the Paterson Institute for Cancer Research, Manchester, U.K.. Preliminary results, which these researchers have kindly allowed us to display here (Figure 5) on chicken, clearly show a substantial difference between unirradiated controls and irradiated samples. This approach seems to have a lot of promise, although the effects of irradiation conditions, post-irradiation storage etc. must be examined.

The use of the radio-isotope ^{32}P in routine testing is undesirable and could be replaced by the use of fluorescent probes, which would be much safer and easier to handle.

Specific Base Modifications

Many radiation-modified bases have been identified and some of these have been suggested as markers for radiation damage. The first difficulty to overcome is that of sensitivity, as already mentioned the yield of any particular product will be very low (at the likely permitted doses <0.1µmol kg^{-1}). A second problem is that the modified base is still incorporated in the DNA which makes it inaccessible to most forms of physico-chemical analysis. Enzymatic hydrolysis, as used in the DNA post-labelling technique can be applied to degrade the DNA to a mixture of nucleotides, these are then amenable to analysis by HPLC. However, detecting the small quantity of modified base is not achievable employing normal optical detection. A radiation product of guanosine, 8-hydroxy-guanosine can be detected electrochemically with a sensitivity several orders of magnitude higher than conventional absorbance detection.[40] Several research groups are involved in looking at this approach as a means of detecting irradiated food.

On subjection to high temperature formic acid hydrolysis some of the modified DNA bases are released from the DNA chain. Their yields and identities can then be determined, after derivatisation, by gas chromatography combined with mass spectrometry.[22] This technique has the great advantage that it provides both a finger-print of the damage as well as identifying specific radiation products, whether it is suitable for the detection of irradiated food is not yet known.

Thymidine glycol is produced by irradiating thymidine containing systems both in the absence and presence of oxygen (see Section 2). On alkaline hydrolysis, thymidine glycol is cleaved open and acetol is released. By radioisotopically labelling the methyl group of the thymidine moiety prior to exposure to irradiation, the amount of acetol and hence the amount of glycol has been quantified.[41,42] More recently this method has been applied to the detection of irradiated foods, the released acetol being reacted with p-amino benzaldehyde to yield a product which can be detected fluorometrically.[43] Appreciable increases in fluorescence have been reported for irradiated (10 kGy) prawns, cod and chicken as compared to unirradiated controls. The reported radiolytic yields seem somewhat larger than might be expected on the basis of the test only determining thymidine glycol. In fact it seems likely that other thymidine products also yield fluorescent material.[33] In addition, in some systems there may be problems associated with high 'background' levels in unirradiated controls.

Immunological assays are widely used in medicine and for food analysis, they offer high specificity combined with high sensitivity and relative ease of application. A number of experimental techniques have been developed for the 'immunological assay' of nucleic acid components.[44] Several research groups have designed such assays for modified DNA bases .[45-47]

Immunological assays rely on the ability of an antibody to bind strongly to a particular antigen. The selected antigen (the product to be analysed) is injected into an animal host which then raises antibodies against it. These antibodies can then be used in assays. The antibodies are present in the animal's blood serum and this can be used directly, alternatively the antibody producing cells can be fused with transformed cells to give hybrid cells which can be grown 'in vitro', thus providing a continuous source of antibodies.

Immunological assays take several forms, e.g. radioimmunoassays (RIA), enzyme linked immunosorbent assays (ELISA). Since the ELISA method would seem most appropriate for the routine analysis of foods (i.e. no radioactive isotopes are necessary) only it will be discussed in any detail. A schematic representation of an ELISA (there are, of course, other variations) is given in Figure 6. The assay involves setting up a competition between antigens in solution and antigens attached to a polystyrene

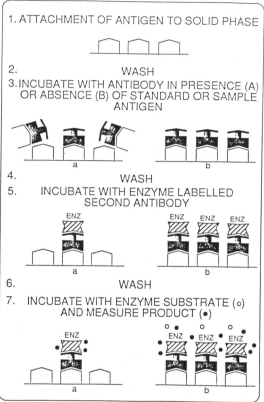

Figure 6. Illustration of the steps involved in an ELISA assay (Z = modified base)

well for binding to a fixed number of antibodies. Proteins bind to polystyrene, consequently when the antigen is a protein or a smaller molecule bound to a protein there is no difficulty in attaching it to the plate. DNA or modified DNA can be attached to the plate by precoating the plate with protamine sulphate to which the DNA then binds.[46] Those antigens in the solution contain the compound to be analysed and may either be authentic standards or derived from the sample under analysis. The antibody-antigen complexes formed free in solution are removed by washing while the antibodies bound to polystyrene-linked antigens remain and can be quantified using a second antibody which has the enzyme horseradish peroxidase (HRP) covalently attached. The amount of enzyme can readily be assessed from its catalysis of the oxidation of 2, 2' azino-bis (3-ethylbenzthiazoline - 6 - sulphonic acid) (ABTS) by hydrogen peroxide, oxidised ABTS absorbs strongly at 405nm and its formation can be monitored optically. Thus, in the absence of competitive antigen, all the antibodies present will bind to the polystyrene-plate-bound-antigen, these antibodies are then complexed by the HRP-linked antibodies which in turn give rise to the formation of a 405nm absorbance on the addition of H_2O_2/ABTS. When free antigen is present in solution fewer antibodies complex with plate-bound antigen which in turn manifests itself as a reduction in 405 nm absorbance. The larger the amount of antigen present in the sample (i.e. free antigen), the larger the inhibition of binding to plate-bound antigen, and the smaller the 405nm absorbance. The technique can be calibrated using known concentrations of authentic antigen, practical aspects of the method are described elsewhere. ELISA's have been developed for the determination of very low concentration of particular compounds, the detection of concentrations in the pico- to femto-mol dm^{-3} range being not uncommon. In the case of thymidine glycol the conversion of one in 10^6 thymines in DNA was sufficient for detection by an immunoassay.[46] Such a high degree of sensitivity should be easily sufficient for the detection of radiation induced glycol in DNA even following doses much lower than that likely to be applied to food. A further advantage of immunoassays is that it is not necessary to degrade the DNA prior to testing, and damaged bases within the DNA could be quantified directly (denaturation of the DNA would probably still be necessary, but this merely requires heating). In the most optimistic scenario, it may not even be necessary to extract the DNA, assays being performed directly on homogenised samples.

The ELISA technique has great potential and in our laboratory antibodies have been raised against thymidine glycol and dihydrothymidine. Following cell fusion antibodies from a number of the resulting hybridomas are currently being tested and the conditions of the assay refined for maximum sensitivity.

A unique characteristic of the chemical effects of radiation is the simultaneous formation of a reducing species (the electron) and an oxidising species (radical cation). Thymidine in irradiated DNA (among other possibilities) can thus be reduced to dihydrothymidine or oxidised to thymidine glycol and the detection of both products is a sure sign of exposure to radiation. Although the glycol is produced in both the presence and absence of oxygen, dihydrothymidine formation is likely to be reduced by the presence of oxygen. Dose rates utilised for food treatment are high, so that oxygen will be removed during irradiation and not replaced by diffusion. Sealed or evacuated samples are frequently treated, therefore it seems quite probable that dihydrothymidine will be produced in the DNA. In contrast to thymidine glycol, or indeed any other hydroxylated base derivative, which is formed by OH radical attack (and OH radicals can be produced by non radiolytically peroxidative reactions), dihydrothymidine is a definite marker for irradiation. The main thrust of our work is now directed at developing an ELISA to detect dihydrothymidine in DNA and to use this as the basis for a conclusive assay for food irradiation.

4 CONCLUDING REMARKS

Most foods contain DNA and the detection of radiation induced changes in DNA has great potential as the basis of an assay for irradiated food. Chain scission and gross changes in structure are relatively simple to detect and might be used to decide whether more detailed analysis is necessary. Alterations to the DNA bases could provide a specific indication of radiation treatment. Appropriate assays are being developed, once these are in operation, the effects of both irradiation and storage conditions can be investigated.

ACKNOWLEDGEMENTS

We are most grateful for a grant from the Ministry of Agriculture, Fisheries and Food which has enabled us to work in this field.

REFERENCES

1. C. von Sonntag, 'The Chemical Basis of Radiation Biology', 1987, Taylor & Francis, London.
2. C. von Sonntag and H.-P. Schuchmann, Int. J. Radiat. Biol. 1986, 49, 1.
3. G. Scholes, 'Effects of Radiation on DNA' (eds. J. Hüttermann, W. Köhnlein, R. Téoule and A. J. Bertinchamps), Springer Verlag, Berlin, 1978, p. 153.
4. D. Schulte-Frohlinde, Adv. Space Res. 1986, 6, 89.
5. J. Ward, Radiat. Res. ,1985, 104, 5.
6. J. D. Watson and F. H. C. Crick, Nature, 1953, 171, 737.

7. A. J. Swallow, 'Radiation Chemistry. An Introduction.', Longman, London, 1973.
8. J. Cadet, A. Balland and M. Berger, Int. J. Radiat. Biol. ,1981, 39 ,119.
9. M. N. Schuchmann, M. Al-Sheikhly, C. von Sonntag, A. Garner and G. Scholes , J. Chem. Soc. Perkin Trans II, 1984, 1777.
10. G. Scholes and R. L. Wilson, Trans. Faraday Soc., 1967, 63 , 2983.
11. G. Scholes, J. F. Ward and J. J. Weiss, J. Mol. Biol., 1960, 2 , 379.
12. J. J. van Hemmen and J. F. Bleichrodt,Radiat. Res., 1971, 46 ,444.
13. G. Hems, Radiat. Res., 1960, 13 , 777.
14. S. Steenken, Chem. Rev., 1989, 89, 503.
15. S. Fujita amd S. Steenken, J. Am. Chem. Soc., 1981, 103 ,2340.
16. D. J. Deeble, D. Schulz and C. von Sonntag, Int. J. Radiat. Biol., 1986 , 49, 915.
17. M. Dizdaroglu, D. Schulte-Frohlinde and C. von Sonntag, Z. Naturforsch., 1975, 30c, 826.
18. D. J. Deeble and C. von Sonntag, Int. J. Radiat. Biol., 1984, 46, 247.
19. D. G. E. Lemaire, E. Bothe, and D. Schulte-Frohlinde, Int. J. Radiat. Biol.,1984, 45, 351.
20. M. N. Schuchmann and C. von Sonntag, J. Chem. Soc. Perkin Trans II, 1983, 1525.
21. R. Téoule, A. Bonicel, C. Bert., J. Cadet and M. Polverelli, Radiat. Res.,1974, 57, 46.
22. A. F. Fuciarelli, B. J. Wegher, E. Ajewski, M. Dizdaroglu and W. F. Blakely, Radiat. Res., 1989, 119, 219.
23. S. Das, D. J. Deeble and C. von Sonntag, Z. Naturforsch., 1984, 40c, 292.
24. D. J. Deeble, S. Das and C. von Sonntag, J. Phys. Chem. 1985, 89, 5784.
25. J. Cadet, A. Balland and M. Berger, Int. J. Radiat. Biol., 1981, 39, 119.
26. P. M. Cullis and M. C. R. Symons, Radiat. Phys. Chem., 1986, 27, 93.
27. J. Hüttermann, K. Voit, H. Oloft, W. Köhnlein, A. Gräslund and A. Rupprecht, Faraday Soc. Discuss., 1984, 67, 135.
28. S. Gregoli, M. Olast and A. Bertinchamps, Radiat. Res., 1982, 89, 238.
29. D. J. Deeble, M. N. Schuchmann, S. Steenken and C. von Sonntag, 1990, J. Phys Chem., in press.
30. K. Murray and W. Old, Prog. Nucleic Acid. Res. and Mol. Biol., 1974, 14, 117.
31. P. H. von Hippel and K. Wong, J. Mol. Biol., 1971, 61, 587.
32. H. Utiyama and P. M. Doty, Biochem., 1971, 10, 1254.
33. A. W. Jabir, D. J. Deeble, P. A. Wheatley, C. J. Smith, P. C. Beaumont and A. J. Swallow, Radiat. Phys. Chem., 1989, 34 , 935.
34. J. Flegeau, M. Copin, C. M. Bourgeois, IJH Heft, 1988, 125, 453.
35. M. A. Harmey, reported at BCR workshop on 'Potential new methods of detection of irradiated food', Committee Bureau of Reference, Commission of the European Communities , Cadarache, France, Feb. 1990.
36. C. Hasselmann and E. Marchion, Ann. Fals. Exper. Chem. Toxicol., 1989, 82 ,169.
37. R. Téoule and J. Cadet, Chem. Commun., 1971, 1269.
38. M. Checko and R. C. Gupta, Carcinogenesis, 1988, 9 , 2309.
39. B. P. Dunn, J. J. Black and A. Naccubbin, Cancer Res., 1987, 47 , 6543.
40. R. A. Floyd, T. A. Watson, P. K. Wong, D. H. Altmiller and R. C. Rickard, Free Rad. Res. Commun. 1986, 1,163.

41. P. Hariharan and P. Cerutti, Proc. Nat. Acad. Sci. (USA)., 1974, 71, 3532.
42. R. L. Warters and J. L. Roti Roti, Int. J. Radiat. Biol., 1978, 34, 381.
43. K. Pfeilsticker and J. Lucas, Angew. Chem., 1987, 99, 341.
44. P. T. Strickland and J. M. Boyle, Prog. Nucleic Acid Res. and Mol. Biol., 1984, 31, 1.
45. G. J. West, I. W. West and J. F. Ward, Radiat. Res., 1982, 90, 595.
46. S. A. Leadon and P. C. Hanawalt, Mutation Res., 1983, 112, 191.
47. R. Rajagopalan, R. J. Melamede, M. F. Lapia, B. F. Erlanger and S. S. Wallace, Radiat. Res., 1984, 97, 499.

Can ESR Spectroscopy be Used to Detect Irradiated Food?

M. H. Stevenson[1,2] and R. Gray[1]

[1]FOOD AND AGRICULTURAL CHEMISTRY RESEARCH DIVISION
DEPARTMENT OF AGRICULTURE FOR NORTHERN IRELAND AND
[2]THE QUEEN'S UNIVERSITY OF BELFAST, NEWFORGE LANE,
BELFAST BT9 5PX, NORTHERN IRELAND

1 INTRODUCTION

The process of irradiation can be well controlled by good management at the irradiation facility and the routine use of dosimeters to measure the dose of ionising radiation absorbed by foods. Nevertheless, it is generally accepted that the development of a test or tests for the detection of irradiated food would facilitate international trade in irradiated food and enhance consumer confidence in the existing control procedures.

Over many years, extensive research programmes have been devoted to understanding the chemical changes which occur in irradiated foods and to establishing the effects of irradiation on the microbiological, organoleptic and nutritional quality of foods. Less effort has been directed towards the development of detection methods. This latter area is now being actively studied and a number of interesting developments including the use of electron spin resonance (ESR) spectroscopy for the detection of irradiated food are being examined.

<u>Characteristics of a Detection Method</u>

Ideally a detection method should be specific for irradiation, that is, no other process should produce the same change in foods. Also, the change induced in the food should be sufficiently large to be measurable and be detectable throughout the expected shelf-life of the food. In some situations, the detection method need only be qualitative while in others it may be desirable or possibly essential for it to be quantitative.

2 SCOPE FOR ESR SPECTROSCOPY

Taking these points into consideration, the usefulness of ESR spectroscopy for the detection of irradiated food will be discussed.

The technique has been applied mainly to foods containing bone but food with shells attached, fruits and vegetables have also been studied.

Food Containing Bone

When bone is subjected to ionising radiation, free radicals are trapped in the crystal lattice of the bone[1,2] and these can be detected using ESR spectroscopy. The technique has been used to date archaeological specimens[3,4] and as an in vivo dosimeter for humans to assess their exposure to radiation.[5]

Ionising radiation generates two predominant paramagnetic species in mineralised tissues at room temperature.[6] The first species which is associated with the organic fraction of the tissue is probably derived mainly from collagen and the second is derived from the crystalline fraction of the bone, the hydroxyapatite.

The paramagnetic species produced in collagen by loss of hydrogen atoms from the amino acid linkages of the protein are characterised by a symmetric doublet with a g value of 2.0032. These collagen radicals are not particularly stable and disappear completely from powdered bone samples in 3 to 6 days if the samples are exposed to oxygen.

The most significant paramagnetic centre is localised in the crystal lattice of hydroxyapatite and gives an asymmetric line characterised by g values at 2.0036 and 1.9978. The identification of the principal ESR signals has proved to be difficult because several radical species are probably trapped in the sample. It has been suggested than one of the trapped species is CO_3^{3-}[7] but contrary to this it has been reported that CO_2^- is more important.[8]

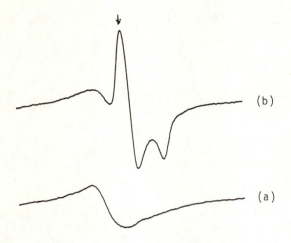

↓ = 3.49 × 10⁻¹ T

Figure 1 ESR spectra of chicken bones (a) unirradiated (b) irradiated at 5 kGy

Free radicals also exist in unirradiated powdered bone samples. A weak broad ESR signal is observed which is further enhanced by such processes as grinding and heating[9] but it is quite different to the large asymmetric singlet induced by ionising radiation (Figure 1). This signal is very characteristic of irradiation and it has been suggested that it would provide a method for the detection of irradiated food containing bone.[9-14] The shape of the signal produced from bones of different species, for example, chicken, turkey, goose, pig, fish and frog legs, is essentially the same and so it can be used as a marker for irradiation in a range of foods containing bone.[12,14-16]

When the only requirement of the method is qualitative detection, the form in which the sample is presented to the ESR spectrometer is not crucial. Thus either fragments of bone or ground bone can be used.

As well as using ESR spectroscopy to qualitatively differentiate between irradiated and unirradiated foods, the feasibility of using the technique to provide a quantitative measure of the dose of

Table 1 Effect of sample preparation method on the ESR signal strength from irradiated chicken bones

Method of preparation	ESR signal strength per g DM
Fragmented	6.889
Fragmented, freeze-dried	9.965
Fragmented, oven-dried	5.188
Microwave-dried, ground	6.643
Freeze-dried, ground	9.390
Oven-dried, ground	5.005
SEM	0.3883
Significance	***

*** $P<0.001$, DM dry matter

(Reprinted with permission from Stevenson, M.H. and Gray, R. J. Sci. Food Agric., 1989, 48, 261. Society of Chemical Industry)

Table 2 Effect of irradiation dose on the ESR signal strength of irradiated chicken drumsticks

Dose/kGy	†Dose Corrected	†Ash Corrected	†Ca Corrected	†P Corrected
2.5	0.87	0.87	0.85	0.88
5.0	1.69	1.70	1.67	1.75
7.5	2.36	2.37	2.36	2.41
10.0	3.09	3.12	3.10	3.16
SEM	0.038	0.038	0.041	0.045
Significance				
Overall	***	***	***	***
Linear	***	***	***	***

*** $P<0.001$, †Values adjusted to standard dose, ash, Ca and P levels.

(Reprinted with permission from Stevenson, M.H. and Gray, R. J. Sci. Food Agric., 1989, 48, 269. Society of Chemical Industry)

irradiation received by a food has been investigated. In this case, the method used to prepare the bone can influence the ESR response, the intensity of which can be determined by either measuring peak height[12] or integrating the area under the absorption curve.[9] Using chicken bones, the effect of different sample preparation methods on the intensity of the ESR signal induced by irradiation has been studied.[9] After double integration of the radiation-induced ESR spectrum (spectrum from an unirradiated control sample having previously been subtracted), and correction of this value for differences in spectrometer tuning, weight of sample in the "active length" of the ESR tube and dry matter concentration, it was concluded that freeze-drying and grinding gave an ESR signal of similar intensity to that of a fragmented, freeze-dried bone (Table 1). The grinding procedure had the advantage of giving a more homogeneous sample. For this reason, ground samples are generally used when quantification of the dose received by bone is attempted.

If ESR spectroscopy is to prove useful for the determination of the irradiation dose received by bone then the ESR signal must show a dose response relationship and be stable throughout the expected shelf-life of the food under the conditions in which it is likely to be stored.

As the irradiation dose increases, the intensity of the ESR integral increases linearly[14] (Table 2) from

Table 3 Effect of length and temperature of storage on the ESR signal strength of irradiated chicken drumsticks

Temp/°C	Duration of storage/days					SEM	(Sig.)
	0	7	14	21	28	S	TxS
	†Dose corrected ESR signal strength						
-20	2.03	2.08	2.05	1.96	2.01	0.043	0.061
+5	2.27	1.87	2.01	1.85	1.92	***	***
Mean	2.15	1.97	2.03	1.90	1.96		

*** $P<0.001$ S = Storage   T = Temperature
† Values adjusted to standard dose values.

(Reprinted with permission from Stevenson, M.H. and Gray, R. J. Sci. Food Agric., 1989, 48, 269. Elsevier Applied Science Publishers)

Table 4 Effect of age of bird on the ESR signal strength of irradiated chicken drumsticks together with the crystallinity ratio

Age/weeks	ESR signal strength		Crystallinity ratio
	[†]Dose corrected	[†]Ash corrected	
8	2.489	2.464	2.60
7	2.323	2.304	2.55
6	2.099	2.107	2.48
5	2.021	2.036	2.42
4	1.913	1.946	2.32
SEM	0.0638	0.0630	0.080
Significance			
Overall	***	***	NS
Linear	***	***	***

NS, not significant, *** $P<0.001$, [†] Values adjusted to standard dose and ash values.

(Reprinted with permission from Gray, R. et al. Radiat. Phys. Chem., 1990, 35, 284. Pergamon Press PLC).

2.5 kGy to the overall average maximum dose (10 kGy) recommended by a number of expert committees.[17,18] Correction of the values to a standard ash, calcium or phosphorus content did not significantly reduce the variability in the results. Even at doses down to 1 kGy, the response has been shown to be proportional to dose.[13] The characteristic radiation-induced ESR signal is quite stable (Table 3) although there is an indication that when irradiated chicken carcasses are stored at +5° C there may be a slight decrease in the intensity of the signal induced in bone stored for up to 28 days.[14]

ESR signal strength is affected by the crystallinity of bone,[19,20] a measurement of which can be obtained by calculating the crystallinity ratio. This is the ratio of the concentration of stable free radicals induced by a saturating dose of ionising radiation to the ash content of the tissue.[19] Using chicken drumsticks (tibia) from birds aged 4, 5, 6, 7 and 8 weeks of age, it has been shown that there is a linear relationship between the age of the bird and the ESR signal strength[20] (Table 4). Regression analysis of the results was carried out to relate the ESR integral to both the age of the chickens and the applied irradiation dose. For each age, there was a linear relationship between ESR intensity and

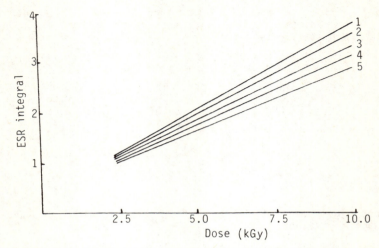

1 = 8 weeks; 2 = 7 weeks; 3 = 6 weeks; 4 = 5 weeks; 5 = 4 weeks

Figure 2   Relationship between ESR integral and irradiation dose for bones from different ages of birds

(Reprinted with permission from Gray, R. et al. Radiat. Phys. Chèm., 1990, 35, 284, Pergamon Press PLC)

irradiation dose over the range 2.5 to 10.0 kGy (Figure 2). A common intercept could be assumed for each age but the regression coefficient increased linearly with age. The crystallinity ratio also increased linearly with age (Table 4), so the change in crystallinity may account for part of the difference in ESR signal strength in bones from birds of different ages. Similar results, although not quantified, have been reported for pork and salmon bones where the radiation-induced signal was considerably more intense from the pork bone than that from salmon bone and using X-ray diffraction it was confirmed that the pork bones were more crystalline.[16]

The influence of the extent of crystallinity on the strength of an ESR signal derived from bone may also be important in quantification of dose received if bones from various sites within a chicken carcass are used. Thus although the shape of the signals obtained from various bones within an irradiated chicken carcass are similar (Figure 3), the signal intensity may be different. Thus tibia and femur bones from chicken

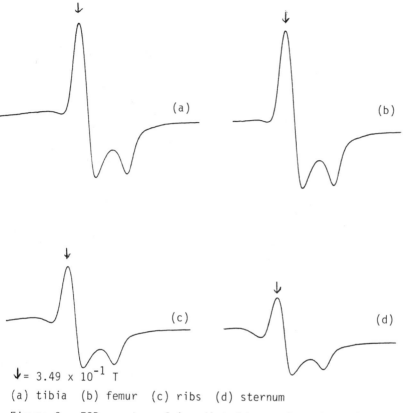

$\downarrow$ = 3.49 x 10^{-1} T
(a) tibia (b) femur (c) ribs (d) sternum

Figure 3 ESR spectra of irradiated bones from four sites within the chicken carcass

carcasses given an irradiation dose of 5 kGy gave radiation-induced ESR signals of similar but significantly higher intensity than those obtained from rib and sternum (Table 5). Again the variations between bones are at least partly due to the differences in bone crystallinity (Table 5).

Other factors which need to be considered before attempting to quantify irradiation dose are the effect of cooking both before and after irradiation. Variable results have been reported in the literature. On the one hand cooking chicken which has been irradiated did

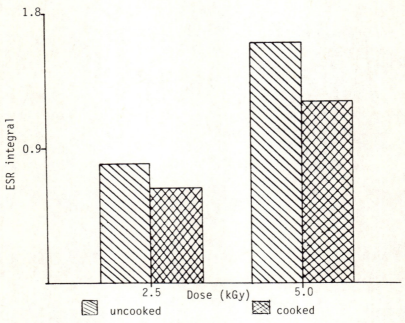

Figure 4 Effect of cooking on the ESR signal strength of bones excised from irradiated chicken drumsticks

(Reproduced with permission from Gray, R. and Stevenson, M.H. Int. J. Food Sci. Technol., 1989, 24, 447. Blackwell Scientific Publications Limited)

Table 5 The ESR signal strength of irradiated tibia, femur, rib and sternum together with their crystallinity ratios

Bone site	ESR values [†]Dose corrected	SEM (Sig)	Crystallinity ratio	SEM (Sig)
Tibia	1.800		1.74	0.043
Femur	1.634	0.1195***	1.59	0.043***
Rib	1.005		1.31	0.075
Sternum	0.845		1.28	0.053

ESR values are the mean of 42 measurements. Crystallinity values are the mean of 18 (tibia and femur), 6 (rib) and 12 (sternum) measurements.

*** $P<0.001$, [†]Values adjusted to standard dose

Can ESR Spectroscopy be Used to Detect Irradiated Food?

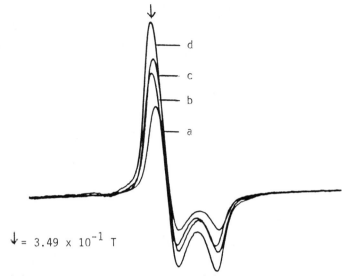

↓ = 3.49 x 10^{-1} T

(a) cooked after irradiation (b) irradiated frozen
(c) irradiated fresh (d) cooked before irradiation

<u>Figure 5</u> ESR spectra of bones excised from irradiated chicken drumsticks

not significantly affect the ESR signal strength[21] while another report recorded approximately a 23% reduction in signal strength (Figure 4).[22] Nevertheless, the signal was still easily detectable but this decrease may affect the accuracy with which the dose received by the uncooked chicken can be quantified. On the other hand, if chicken is cooked before irradiation, the intensity of the radiation-induced signal increases.[12] Davidson[23] has suggested that because the water content of cooked bones is reduced, there is less opportunity for free electrons to react with water. As a consequence, more free radicals are trapped in bone and give an enhanced ESR signal. In a recent experiment, the differences in ESR signal strength of bones from irradiated fresh, frozen and cooked chicken carcasses has been illustrated (Figure 5). This again highlights how the past history of the bone may affect the response to irradiation.

The use of re-irradiation of bone samples has been suggested as a way of eliminating the necessity of having a detailed knowledge about the history of samples prior to quantification of dose. This procedure has been used to quantify the dose received by chicken, pork and fish bones.[21] Although this is obviously a possible way of dealing with the problem of quantifying dose, it does require the use of an irradiator as well as an ESR spectrometer and also because handling procedures such as cooking can influence the response obtained, the accuracy of the result may be less than legally acceptable. An alternative approach might be to try to develop a relationship between the strength of the ESR signal and irradiation dose incorporating the factors which may influence the response obtained. This type of approach has yet to be tested in practice.

Finally, ESR spectroscopy has been used successfully to differentiate between two unidentified blocks of frozen deboned turkey meat, one of which had been irradiated using a linear accelerator.[24] Extraction of small fragments of bone from both blocks of meat by digestion in 1.0 M alcoholic KOH gave two samples, one of which displayed the characteristic radiation-induced ESR signal. After subjecting sub-samples from the unirradiated block to either 3, 5 or 7 kGy doses of irradiation, a relationship between ESR signal intensity and the dose applied was derived. This relationship was used to estimate the dose received by the deboned meat irradiated using the electron source. The average irradiation dose determined was 4.7 kGy and this was in close agreement with the overall average dose of 5 kGy recorded by the supplier. The fact that the relationship between irradiation dose and signal strength derived using a gamma radiation source (low dose rate) could be used to quantify the dose received by a product treated using a source of high energy electrons (high dose rate) would indicate that even when ionising radiation is applied at very different dose rates, the intensity of the radiation-induced signal is similar.

Food Containing Shells

Initial studies have been carried out on the use of ESR spectroscopy for the detection of irradiated shrimp[11,25] and molluscs.[25] The ESR spectrum derived from the shell of unirradiated shrimp was different to that from irradiated shell. In the latter case there was a sharp singlet at g = 2.005, together with several

Can ESR Spectroscopy be Used to Detect Irradiated Food?

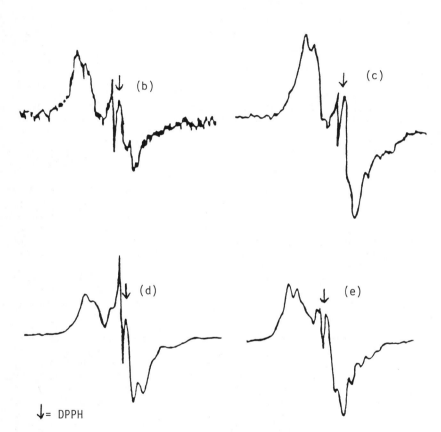

↓ = DPPH

(a) unirradiated (b) 1 kGy (c) 5 kGy (d) 10 kGy (e) 20 kGy

Figure 6   ESR spectra from shells of unirradiated and irradiated shrimps

(Reprinted with permission from Desrosiers, M.F. J. Agric. Food Chem., 1989, 37, 96. American Chemical Society)

other resonances (Figures 6a, b).[25] As irradiation dose increased from 5 to 20 kGy (Figures 6c to e), this radical persisted but there were also a number of other resonances which must have been derived from another radical or radicals. The radical with g value = 2.005 did not increase linearly with dose and the situation was further complicated by the dose dependent changes in the other spectral lines. The reason for these complex changes is not understood and requires further investigation.

The shrimp exoskeleton is composed of chitin, a major component of which is D-glucosamine. When N-acetyl-D-glucosamine was irradiated at 5 kGy in powder form it gave an ESR signal comparable to that obtained from irradiated shrimp shell, thus suggesting that the more complex signal in the shell may be derived from chitin.[25] However, other components of the shell may also contribute to this radiation-induced signal.[25]

When the intensity of the signal was determined by integration, the dose response relationship was linear between 1 and 3 kGy, the dose range likely to be used in practice. Monitoring of the signal induced by a 1 kGy dose of irradiation for up to 43 days revealed no significant decay. However, because of the complexity of the signal at these low doses, it would be advantageous if a method could be developed which eliminated one or more of these signals to leave a single signal which could be more accurately quantified.[25]

This work also highlighted the differences between shrimps from different origins since an earlier report[11] had given ESR signals from irradiated shrimps which were substantially different to those reported by Desrosiers.[25] There is a need for research into the origin of these signals and the factors which influence the responses obtained before attempting to produce a standard procedure which would ultimately have to stand up to the rigours of legal scrutiny.

The ESR signals produced in mussel shells given an irradiation dose of 1 kGy were about 1000 times more intense than those produced in shells from shrimps. The induced signals contain resonances which correlate with radiation-induced signals previously observed in shell[26] and are reported to be due to CO_2^-.[8]

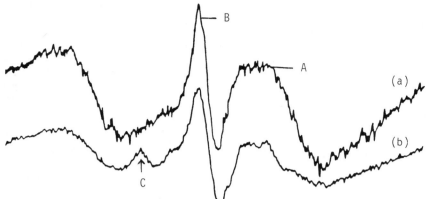

Figure 7   ESR spectra of unirradiated (a) and irradiated (b) strawberries

(Reprinted with permission from Raffi et al. J. Chem Soc. Faraday Trans. 1, 1988, 84, 3359. Royal Society of Chemistry)

Fruits and Vegetables

Recently, a method has been proposed for the identification of irradiated strawberries. The complex spectrum obtained is composed of three signals A, B and C (Figure 7).[27]

Signal A is due to Mn^{2+} and is present in both irradiated and unirradiated samples.[11]

Signal B is a single line present in both irradiated and unirradiated samples. Its origin is not known. The height of the signal increases with irradiation but varies with water content[28] and consequently it would not provide a definitive test of irradiation.

Signal C is only present in irradiated samples and has been shown to be a triplet radical induced in the cellulosic parts of fruits.[28] It would be expected to be present in all fruits, especially in the achenes, pips or stones. A number of other fruits and vegetables have been examined including raspberries, blackberries, grapes, apples, peaches and onion.[16,28,29] Although they have all shown similar signals, the difference between the irradiated and unirradiated samples is not sufficiently large in some instances for the method to prove useful as a definitive test of irradiation. Also, the signal may not persist throughout the storage life of the product. In conclusion, the technique may be useful for detecting irradiated strawberries, raspberries, red currants, bilberries,[28] blackberry, pistachio and onion.[29] In all cases, the method at least at present, will only give a qualitative indication of irradiation. More research is needed to establish the influence of irradiation and processing variables on these signals induced in fruit and vegetables before the potential of the method as a means of detection can be critically evaluated.

3 CONCLUSIONS

ESR spectroscopy does offer potential for the qualitative detection of irradiation in a number of foods, particularly those containing bone. It also shows promise for the detection of irradiated shell fish and may be useful as a screening method for fruit but other methods would need to be developed for these latter foods.

With regard to quantification, the technique may also be able to give an acceptable estimate of dose in foods containing bone but that will have to be tested further by interlaboratory comparisons. The application of the method for quantification of dose in other foods is less well advanced.

The chemist has a fundamental role to play in elucidating the nature of the radicals induced by irradiation and the factors which influence their formation and stability. This knowledge will contribute to the understanding of the changes which occur in irradiated food and help to establish the applicability of ESR spectroscopy as a technique for the detection of irradiation in a range of foods.

REFERENCES

1. W. Gordy, W.B. Ard and H. Shields, Proc. Nat. Acad. Sci. Wash., 1955, 41, 983.

2. K. Ostrowski, A. Dziedzic-Goclawska, W. Stachowicz and J. Michalik, Ann. N.Y. Acad. Sci., 1974, 238, 186.

3. M. Ikeya and T. Miki, Science, 1980, 207, 977.
4. S. Mascarenkas, O. Baffa Filho and M. Ikeya, Amer. J. Phys. Anthropol., 1982, 59, 413.
5. B. Pass and J.E. Aldrich, AAMD Journal, 1986, 11, 7.
6. K. Ostrowski, A. Dziedzic-Goclawska and W. Stachowicz. "Free radicals in biology". Academic Press, London, 1980, Vol 4, Chapter 10, p. 321.
7. R.A. Serway and S.A. Marshall, J. Chem. Phys., 1967, 46, 1949.
8. M. Geoffroy and H.J. Tochon-Danguy, Calcif. Tissue Int., 1982, 34, S99.
9. M.H. Stevenson and R. Gray, J. Sci. Food Agric., 1989, 48, 261.
10. D. Onderdelinden and L. Strachee. "The identification of irradiated foodstuffs". Commission of the European Communities International Colloquim, 1973, 127.
11. N.J.F. Dodd, A.J. Swallow and F.J. Ley, Radiat. Phys. Chem., 1985, 26, 451.
12. J.S. Lea, N.J.F. Dodd and A.J. Swallow, Int. J. Food Sci. Technol., 1988, 23, 625.
13. M.F. Desrosiers and M.G. Simic. J. Agric. Food Chem., 1988, 36, 601.
14. M.H. Stevenson and R. Gray, J. Sci. Food Agric., 1989, 48, 269.
15. J. Raffi, J.C. Evans, J-P. Agnel, C.C. Rowlands and G. Lesgards. J. Appl. Radiat. Isot., 1989, 40, 1215.
16. B.A. Goodman, D.B. McPhail and D.M.L. Duthie, J. Sci. Food Agric., 1989, 47, 101.
17. Joint FAO/IAEA/WHO Expert Committee on the Wholesomeness of Irradiated Food (JECFI), 1981. Report of the Working Party on Irradiation of Food. WHO Technical Report Series No 659.

18. Advisory Committee on Irradiated and Novel Foods (ACINF), 1986. Report on the safety and wholesomeness of irradiated food, HMSO, London.

19. K. Ostrowski, A. Dziedzic-Goclawska, W. Stachowicz and J. Michalik, Basic Appl. Histochem., 1981, 25, 79.

20. R. Gray, M.H. Stevenson and D.J. Kilpatrick, Radiat. Phys. Chem., 1990, 35, 284.

21. N.J.F. Dodd, J.S. Lea and A.J. Swallow, Nature, 1988, 334, 387.

22. R. Gray and M.H. Stevenson, Int. J. Food Sci. Technol., 1989, 24, 447.

23. I.G. Davidson, PhD Thesis, ESR Studies on -irradiated foods, University of Aberdeen, Scotland.

24. R. Gray and M.H. Stevenson, Radiat. Phys. Chem., 1989, 34, 899.

25. M.F. Desrosiers, J. Agric. Food Chem., 1989, 37, 96.

26. M. Ikeya, J. Miyajima and S. Okajima, Jpn. J. Appl. Phys., 1984, 23, L697.

27. J.J. Raffi, J-P.L. Agnel, L.A. Buscarlet and C.C. Martin, J. Chem. Soc. Faraday Trans. 1, 1988, 84, 3359.

28. J.J. Raffi and J-P.L. Agnel, Radiat. Phys. Chem., 1989, 34, 891.

29. M.F. Desrosiers and W.L. McLaughlin. Radiat. Phys. Chem., 1989, 34, 895.

Radiation, Micro-organisms and Radiation Resistance

B. E. B. Moseley

AFRC INSTITUTE OF FOOD RESEARCH, READING LABORATORY, SHINFIELD, READING RG2 9AT, UK

1 INTRODUCTION

Ionizing radiations are mutagenic for most bacteria and are lethal for all of them. For the vast majority of bacteria the critical target for inactivation is the chromosome, a single, circular molecule of DNA containing several million base pairs. Of course, if the sensitivity of bacteria to ionizing radiation was solely the result of chromosome damage it might be expected that most bacterial species would be inactivated by similar doses of radiation since there is not a great variation in the sizes of bacterial chromosomes, by far the majority being in the range of 1.5 to 4 x 10^9 daltons.[1] However, from the time they evolved, all organisms have been exposed to ionizing radiations as part of their environment, both from cosmic rays and radiations emanating from the decay of radioactive materials in rock, soil and biological material.[2] It is not surprising therefore that all organisms have evolved enzymic mechanisms for keeping their DNA in good repair in the face of this environmental onslaught. Thus, although the chromosomes of bacteria are intrinsically very sensitive to potentially lethal damage as a result of exposure to ionizing radiation, the ability of bacteria to repair a limited amount of such damage gives them considerable greater resistance to such radiations than would otherwise be the case. The efficiency with which different bacteria repair the radiation-induced damage to their DNA varies considerably, so species of the most sensitive vegetative bacterial genera, e.g. Pseudomonas, and species of the most resistant, Deinococcus, vary in their resistance by a factor of about 100.

Bacterial endospores (spores) are more resistant to the lethal action of ionizing radiation than their corresponding vegetative cells by a factor of about 5 to 15[3] and in general spores are much more resistant than vegetative bacteria regardless of species.[4] However, an exception to this rule is found in the vegetative cells of <u>Deinococcus</u> species which are more resistant than bacterial spores. These observations are important in considering the value of ionizing irradiation as a means of significantly reducing the numbers of spoilage and/or pathogenic bacteria in food.

2 IDENTIFICATION AND ASSAY OF RADIATION-INDUCED DAMAGE

Evidence for DNA being the Target for Radiation Inactivation

There are good theoretical reasons for believing that the DNA of a bacterial cell is the target for inactivation. Because of the non-discriminating nature of ionizing radiation, all components of the cell absorb radiation and are damaged in proportion to their mass. If a dose of radiation is applied to a cell that causes one chemical lesion per 10^7 molecular weight (M_r), then if a cell has 1000 molecules of an enzyme of say 10^5 M_r, only 10 of the enzyme molecules will be damaged or, put another way, 990 enzyme molecules will remain undamaged. However, a chromosome of 2 to 3 x 10^9 M_r will sustain between 200 and 300 chemical lesions and this may be sufficient to inhibit completely its replication or its ability to be transcribed.

However, the most convincing experimental evidence comes from the isolation of mutants of wild-type bacteria that are more sensitive to ionizing radiation than the parent. A large number of such mutants have been isolated and all appear to be defective in their ability to repair ionizing radiation-induced lesions in DNA.[5]

Since ionizing radiation may deposit energy in any of the atoms that make up DNA, the number of alterations that can occur in DNA runs into hundreds. Many of these have been identified by irradiating free bases, deoxyribonucleosides, deoxyribonucleotides and short oligonucleotides and examining the products, but their relevance to DNA irradiated in the bacterial cell may be minimal. For example, some products appear only at doses that are not biologically significant, i.e. well above the dose required to inactivate bacteria, and thus occur only in the DNA of bacteria that are already non-viable. However, some forms of damage have been closely correlated with cell killing while some others have been implicated. Thus the formation of single-strand (s-s) and double-strand (d-s) breaks has been extensively studied. Evidence for relevant base damage is accumulating but its role in cell killing is still uncertain. For reviews of ionizing radiation-induced damage to DNA, see references 6-9.

Strand Breakage

DNA strand breakage occurs following predominantly indirect action on the deoxyribose moiety of the sugar-phosphate backbone of the DNA. About 20% of the hydroxyl radical reactions with DNA result in the abstraction of hydrogen atoms from any of the five carbon atoms of the deoxyribose moiety and subsequent β-elimination leads to strand breaks usually from scission of the C-3' phosphate ester bond. The majority of strand breaks originate from cleavage of this bond so that $5'-PO_4$ and $3'-PO_4$ termini are found in a 3:1 ratio. In the former case there is an altered sugar on the 3' terminus and in half of these the damaged 3' nucleoside end group splits off leaving a small gap with a $5'-PO_4$ and $3'-PO_4$ bordering it.

The introduction of strand breaks into bacterial DNA during ionizing irradiation can be followed by measuring the sedimentation velocity of irradiated DNA and comparing it with that of DNA from unirradiated bacteria. This method was developed by McGrath & Williams who showed that it was possible to isolate and sediment relatively high molecular weight single-strand DNA from bacteria by lysing cells on top of an alkaline 5% to 20% sucrose gradient.[10] As the DNA accumulates breaks, the average molecular weight of the DNA is reduced and this is reflected by a slower rate of sedimentation. There are many measurements of strand breaks in alkali in DNA from irradiated bacterial cells. Most fall in the range of 1 to 10×10^{-12} strand breaks in alkali per rad (10^{-2} Gy) per dalton.[9] Some of the variation is due to experimental error and it is reasonable to take the value of 5×10^{-12} strand breaks in alkali per rad per dalton as the yield in oxygen, which is about four times as many as when irradiated under anoxic conditions and ten times as many as when irradiated in the dry state.

Certain minor radiation products of deoxyribose do not lead directly to strand breakage in the cell but do so on treatment with alkali. About one-third of the s-s breaks detected by alkaline sucrose gradient techniques arise from such alkali-labile sites.[11]

A DNA d-s break is formed when the two strands of the double helix are broken opposite each other or at least sufficiently close to each other that the hydrogen-bonding between the two sites is insufficient to maintain the structure. Double-strand breaks can be detected by measuring the sedimentation velocity of DNA from irradiated bacteria in neutral, rather than alkaline, sucrose gradients and comparing it with that of DNA from unirradiated bacteria.[12] The number of d-s breaks in irradiated DNA is about 20-fold less than the number of s-s breaks. At low doses of ionizing radiation the number of d-s breaks increases linearly with dose, implying that the two breaks are caused by a single absorption event. Of course, as the irradiation dose increases, then the chance of two independently created s-s breaks forming a d-s break by coincidence increases with the square of the dose.

Base Damage

Most damage to DNA irradiated in cells is caused by the hydroxyl radical, a radiolytic product of water. About 20% of this attack is on DNA sugars and nearly 80% on bases, with the sensitivity of the bases being in the order of thymine, cytosine, adenine, guanine. A variety of radiolysis products of pyrimidines has been detected but the major product is formed by hydroxyl radical attack on the 5,6 double bond, with both the C5 and C6 being equally susceptible, and resulting in the formation of lesions of the 5,6 dihydroxy-5,6-dihydrothymine type. Such lesions were first recognized as ionizing radiation-induced products of thymine in Deinococcus radiodurans and mammalian cells.[13,14] In D.radiodurans the amount of this thymine derivative formed in the DNA was calculated to be 1.2×10^{-12} per dalton per rad (10^{-2} Gy). Other products of hydroxyl attack on bases have been detected in the DNA of irradiated cells, e.g. cis thymine glycols, 5-hydroxymethyl uracil (see reference 9, for yields etc.), but the relevance of all these products to cell death is not clear.

3 REPAIR OF RADIATION-DAMAGED DNA IN BACTERIA

Survival Curves and the Extent of DNA Damage

The sensitivity of a population of bacteria to the lethal effects of ionizing radiation is usually depicted in the form of a survival curve in which the surviving fraction of an irradiated population is plotted on a logarithmic scale against dose on a linear scale (Figure 1).

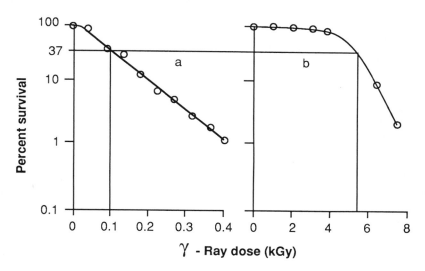

Figure 1 Survival curves of γ-irradiated (a) <u>Escherichia coli</u> B/r and (b) <u>Deinococcus radiodurans</u>. The D_{37} value, i.e. the dose which approximates to that required to inactivate a single viable unit, is indicated. The values for <u>E.coli</u> and <u>D.radiodurans</u> are 0.1 and 5.5 kGy respectively and may be used to calculate the number of DNA lesions per genome at the lethal dose. Data from reference 15.

Because of the random nature of radiation damage to DNA, potentially lethal DNA lesions are scattered randomly through the population, i.e. they follow a Poisson distribution; at a radiation dose which theoretically would kill all the bacteria were it evenly distributed between them, 37% of them will survive, having received less than an inactivating dose, while some of the 63% killed will have received a dose in excess of that required to inactivate them. The survival curves of the majority of bacterial species are similar to those shown in Figure 1(a), i.e. they have an extremely small 'shoulder' followed by an exponential loss of viability. Some vegetative bacteria and spore populations, however, have survival curves of the type shown in Figure 1(b) in which there is a very large shoulder before exponential loss of viability occurs. In these cells a substantial amount of sub-lethal damage is accumulated before inactivation occurs. The absence of a shoulder on a survival curve should not be taken to mean an absence of a repair mechanism for potential lethal radiation damage in the bacterium under study – a bacterial mutant with no repair mechanism for ionizing radiation damage is extremely sensitive – but that the lesion which inactivates the cell is relatively rare although it is formed linearly with dose. In contrast, the reasons why cells are inactivated exponentially after a long period of accumulating sub-lethal damage are complex and probably represent the difficulty of repairing a large number of similar lesions some of which are clustered in the DNA.

Whatever the explanation, the radiation dose that kills 63% of the initial population of bacteria or, put another way, allows 37% to survive (the D_{37}) is taken as the dose required to kill a single cell and its value can be derived from the survival curve (Figure 1). Thus the D_{37} for <u>Escherichia coli</u> B/r is 10 krad (0.1 kGy) while for <u>D.radiodurans</u> irradiated under identical conditions it is 550 krad (5.5 kGy), i.e. <u>D.radiodurans</u> is 55 times more resistant than <u>E.coli</u> B/r. Since the rate at which s-s and d-s breaks and base damage are introduced into DNA by irradiation is known, and the sizes of the <u>E.coli</u> and <u>D.radiodurans</u> chromosomes are known, it is possible to calculate the numbers of each kind of lesion in the respective chromosomes at the lethal dose and it is assumed that one of these forms of damage is responsible for cell death, the most likely candidate in <u>E.coli</u> being the d-s break.

The estimates derived in this way for <u>E.coli</u> show some variation but approximate values at the D_{37} would be 80 to 160 s-s breaks per genome, 4 to 6 d-s breaks and 20 to 40 products of the 5,6-dihydroxy-5,6-dihydrothymine type. The variation depends on the value of the D_{37} which can vary, for example, with the strain of <u>E.coli</u> irradiated and with the physiological state of the culture, i.e. whether exponentially growing or resting phase cells are irradiated, the latter being more resistant than the former. For <u>D.radiodurans</u> the corresponding values would be about 3000 s-s breaks, 150 d-s breaks and 1000 5,6-dihydroxy-5,6-dihydrothymine products. The majority of vegetative bacteria are more likely to be in the same range as <u>E.coli</u> than <u>D.radiodurans</u>.[16]

Comparative Resistance of Bacteria to Ionizing Radiation

The proportion of a bacterial population that survives a particular dose of radiation will depend on the intrinsic sensitivity of the species, its stage during the growth cycle, the amount of radiation damage inflicted, its potential for repair and the relative proportions of damage channelled into different repair pathways.

The maximum sensitivity of a species would be exhibited when it is harvested in rapid exponential-phase growth, resuspended in phosphate buffer, irradiated while bubbling oxygen through the suspension and then plated on a rich medium. Thus resting-phase bacteria are more resistant to ionizing radiation than those in exponential phase; bacteria appear to be more resistant when irradiated in a broth medium than in phosphate buffer; freeze dried bacteria are more resistant than normal "wet" bacteria; bacteria irradiated in the absence of oxygen are more resistant than those irradiated in its presence. As an example, <u>D.radiodurans</u> irradiated in phosphate buffer, broth and as a freeze-dried culture requires doses of 7.5, 13.2 and 64 kGy respectively to reduce its viability by 90%.

If bacterial species are compared under standard conditions, in general Gram-negative organisms such as <u>Salmonella</u> and <u>Campylobacter</u> are more sensitive to radiation than Gram-positive ones such as <u>Listeria</u> and <u>Staphylococcus</u> while bacterial spores are much more resistant. Spores belong mainly to two groups of bacteria of great interest to food microbiologists; the <u>Bacillus</u> and <u>Clostridium</u> genera. Even more resistant than these spores are the vegetative cells of the <u>Deinococcus</u> and <u>Deinobacter</u> genera but fortunately these are not pathogenic and are extremely sensitive to heating.

4 SELECTION AND INDUCTION OF RADIATION-RESISTANT BACTERIA

The Selection of Radiation-Resistant Strains

Clearly the microbiological flora of most food items suitable for irradiation will consist of a mixture of genera and species of lesser or greater resistance to irradiation. Since the purpose of food irradiation is to reduce substantially the numbers of contaminating spoilage bacteria and to eliminate the serious food-borne pathogens such as <u>Salmonella</u>, <u>Campylobacter</u> and <u>Listeria</u> which between them cause more than 90% of deaths from bacterial food poisoning, there is inevitably a selection for the more radiation-resistant bacteria in the surviving population.

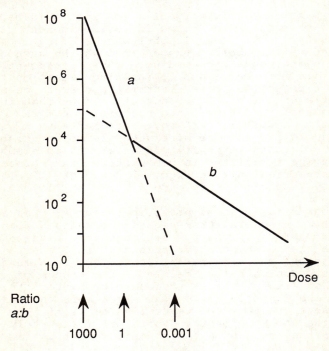

<u>Figure 2</u> Survival curve for a mixed population of bacteria. There are two components, a relatively radiation-sensitive one present at 10^8 initially (a) and a relatively radiation-resistant one present at 10^5 (b). For explanation see text.

Figure 2 shows what happens when a mixed population, of two organisms of different sensitivities is irradiated. At time zero the radiation sensitive strain "a" is present at 1000 times the number of strain "b" but as the radiation is applied the ratio becomes 1:1 and then 1:1000 in favour of "b". Thus the contaminating flora of an irradiated product will tend to be Gram-positive rather than Gram-negative and in this sense it is very similar to heat pasteurisation of milk which selects for a heat-resistant population.

It is true that if a food is contaminated with both vegetative spoilage bacteria and spores of <u>Clostridium botulinum</u> that irradiation will destroy the majority of the spoilage organisms but not the spores. The latter would be able to germinate under appropriate conditions and give rise to a population of cells with the consequent formation of toxin. In practice the problem of survival of <u>Clostridium botulinum</u> spores occurs with any non-sterilizing food process so that food irradiation does not pose a unique hazard in this respect. Many heated foods on sale today have been treated to kill vegetative cells but spores, including those of <u>Cl.botulinum</u>, survive. Of course, if a toxin has been produced in a food prior to irradiation it cannot be removed by irradiation even though the microbe that produced it can be. However, it should be remembered that more than 90% of reported cases of food poisoning in the UK (and in the world) do not involve bacteria that produce their effect by producing an extracellular toxin. Additionally, it is envisaged that irradiation will be used to treat fresh food to eliminate most food-borne pathogens and that no special hazard will arise that cannot be satisfactorily evaluated and avoided using standard microbiological and technological methods. The application of such procedures has meant that there has been only one death from botulism as a result of eating food processed in Britain in the last 60 years.

<u>Induction and Selection of Radiation-Resistant Mutants of Bacteria</u>

There is a general worry that the use of ionizing irradiation as a food treatment process will lead to the induction of mutations in contaminating bacteria and to the development of radiation resistance in survivors. In practice it is extremely difficult to induce such mutation especially by a single irradiation treatment.

The only successful method for elucidating biochemical pathways for the repair of radiation-induced DNA damage in bacteria has been to isolate mutants that are more or less resistant to radiation than the wild type or parent strain and then to compare the mutants and wild type biochemically. Technically, the isolation of radiation-resistant mutants is easy, in that if a dose of radiation is applied to a wild-type population so that only a small fraction survives, spontaneous and induced radiation-resistant mutants should be among the survivors. In spite of the technical ease, very few resistant mutants have been isolated. An exception is the strain E.coli B/r isolated from the wild type E.coli B.[17] However, this mutant has approximately the same resistance as the wild type K12 strain, E.coli AB1157. This almost total lack of success in isolating radiation-resistant mutants from a wild type population suggests that wild types or natural strains of bacteria have already evolved an adequate DNA repair capacity. This augurs well for the use of ionizing radiation as a "pasteurizing" treatment for food and food ingredients in which there is a single radiation treatment.

When bacteria are subjected to many cycles of irradiation then it is possible to develop radiation-resistant populations. In these experiments survivors from a single irradiation treatment are grown into a large population and re-irradiated. The survivors from this treatment are grown up again and irradiated, etc. After many such rounds of treatment the population as a whole has, on occasions, been shown to be more resistant than the original one, but usually no genetic analysis has been carried out on the resulting population. An exception is an analysis of a radiation-resistant strain of Salmonella typhimurium isolated as above[18,19] which was shown to be mutant in two genes compared with its respective wild type.[20] However in an irradiation plant, food items would necessarily be enclosed in wrapping to prevent reinfection and would be irradiated only once so the radiation-resistant organisms would not be selected. Radiation, of course, is not unique in this respect; heat-resistant mutants of Salmonella can be isolated after many cycles of exposure to heat[21] but this has not created problems for pasteurizing plants.

5 CONCLUSION

The identification of the lethal and mutagenic lesions induced by ionizing radiation and their repair in bacteria is running some way behind the corresponding study of UV-induced damage. The reasons are not difficult to deduce. The use of germicidal lamps as a source of 254 nm light, the wavelength that is maximally absorbed by nucleic acids, is routine in microbiological laboratories, the lamps are cheap and safety procedures minimal. Laboratory experiments using ionizing radiations require the use of 'hotspot' ^{60}Co γ-ray sources, in which samples to be irradiated are lowered via a piston into a lead housing containing a ring of ^{60}Co rods. These machines weigh several tons and are very expensive, about 100 times the cost of a UV-lamp. Also, the decline over the last 20 years of commercial interest in using ionizing irradiation as a food treatment process has removed considerable financial support for such research.

Although the irradiation of food and food products for human consumption is prohibited in the UK at the time of writing, the general climate of international opinion, and the fact that certain countries already have, or are considering adopting, a more flexible attitude to food irradiation, may cause legislation here and elsewhere to be relaxed in the very near future and research programmes to begin so that the molecular nature of irradiation damage to contaminating bacteria, under a variety of conditions, and their ability to repair such damage and survive will rest on a much firmer base of scientific knowledge.

REFERENCES

1. M. Herdman, 'The Evaluation of Genome Size', ed. T. Cavalier-Smith, Wiley, Chichester, 1985, p. 37.
2. A. Nasim and A.P. James, 'Microbial Life in Extreme Environments', ed. D.J. Kushner, Academic Press, London and New York, 1978, p. 409.
3. C. Woese, J. Bacteriol., 1958, 75, 5.
4. B.A. Bridges, Prog. in Indust. Microbiol., 1964, 5, 283.
5. B.E.B. Moseley and E. Williams, 'Advances in Microb. Physiol.', eds. A.H. Rose and D. Tempest, Academic Press, London and New York, 1977, Vol. 16, p. 99.

6. R.B. Setlow and J.K. Setlow, Ann. Rev. Biophys. Bioengineering, 1972, 1, 293.
7. P.A. Cerutti, 'Molecular Mechanisms for Repair of DNA', eds. P.C. Hanawalt and R.B. Setlow, Plenum Press, New York and London, 1975, p. 3.
8. J. Huttermann, W. Kohnlein and R. Teoule, eds. 'Effects of Ionizing Radiation on DNA', Springer-Verlag, Berlin, 1978.
9. F. Hutchinson, Prog. Nucleic Acid Mol. Biol., 1985, 32, 115.
10. R.A. McGrath and R.W. Williams, Nature, 1966, 212, 534.
11. M.C. Paterson, K.J. Roozen and J.K. Setlow, Internat. J. Radiat. Biol., 1973, 23, 495.
12. H.S. Kaplan, Proc. Natl. Acad. Sci. U.S.A., 1966, 55, 1442.
13. P.V. Hariharan and P.A. Cerutti, Nature, New Biol., 1971, 229, 247.
14. P.V. Hariharan and P.A. Cerutti, J. Mol. Biol., 1972, 66, 65.
15. D.M. Sweet and B.E.B. Moseley, Mutat. Res., 1976, 34, 175.
16. B.E.B. Moseley, 'Mechanisms of Action of Food Preservation Procedures', ed. G.W. Gould, Elsevier Applied Science, London and New York, 1979, Chapter 3, p. 43.
17. E.M. Witkin, Genetics, 1947, 32, 221.
18. R. Davies and A.J. Sinskey, J. Bacteriol., 1973, 113, 133.
19. R. Davies A.J. Sinskey and D. Botstein, J. Bacteriol., 1973, 114, 357.
20. S.N. Ibe, A.J. Sinskey and D. Botstein, J. Bacteriol., 1982, 152, 260.
21. J.E.L. Corry and T.A. Roberts, J. Appl. Bacteriol., 1970, 33, 733.

Dosimetry for Food Irradiation

P. H. G. Sharpe

NATIONAL PHYSICAL LABORATORY, TEDDINGTON, TW11 0LW, UK

1. INTRODUCTION

Food irradiation is just one aspect of the more general field of **radiation processing** in which ionising radiation is used to effect some physical, chemical or biological change in materials. In particular, the techniques of food irradiation have much in common with those of the well established radiation sterilization of medical devices, albeit at lower dose levels, and many of the systems and techniques mentioned in this paper are applicable to either activity.

The current upsurge in interest in food irradiation stems from the 1980 report of the Joint FAO/IAEA/WHO Expert Committee on the Wholesomeness of Irradiated Foods (JECFI) which stated that the irradiation of any food to an overall average dose of 10 kGy causes no toxicological hazard and introduces no special nutritional or microbiological problems. In order to eliminate the possibility of significant induction of radioactivity in food, JECFI further recommended that radiation sources be limited to energies of 5 MeV or less for gamma rays or X-rays, and to 10 MeV or less for electrons. Again, these radiation qualities are the same as have been used successfully for the sterilization of medical devices for the past thirty years.

It will be clear from other contributions to these proceedings that a vital component of any regulatory framework for the control of food irradiation is the ability to make accurate dose measurements that are traceable to appropriate national standards. The aim of this paper is to describe some of the ways in which this can be achieved.

2. FUNDAMENTALS

Before looking in detail at dosimetry systems it is necessary to consider a few fundamental physical aspects of radiation absorption and the implications that these have for food irradiation.

Although photons and high energy electrons are very different forms of radiation with very different penetrative properties (see below), the effect of photons incident on low atomic number materials (eg carbon, hydrogen and oxygen) is to produce a cascade of secondary electrons in the material and it is these secondary electrons, rather than the primary photons, that transfer most of the energy to the absorber. Thus it is possible to regard photons merely as a means of introducing electrons into a medium and, in the energy ranges considered here, both types of radiation produce essentially identical initial chemical effects. This is not to say that the observed effects of electrons and photons will be the same; dose rates from electron beam sources are generally many orders of magnitude higher than from photon sources and dose rate dependent chemical reactions can cause differences in the overall effects.

Absorbed Dose

The quantity of interest in radiation processing is the amount of energy absorbed by the material being treated; radiation which passes through material but does not deposit energy can have no effect! The International Commission on Radiation Units (ICRU) have defined the quantity **absorbed dose** as "the mean energy imparted by ionising radiation to the matter in a volume element divided by the mass of the matter in that volume element". Thus absorbed dose has the units of energy per unit mass and in the SI system has been given the special unit **gray** (Gy) defined as 1 joule per kilogram (1 J/kg). An older unit is the **rad**, defined as 100 erg/g, which although officially discontinued in 1986 still finds use in many establishments.

$$1 \text{ Gy} = 100 \text{ rad} \qquad 1 \text{ Mrad} = 10 \text{ kGy}$$

An important consequence of the definition of absorbed dose is that the quantity refers to energy absorbed in a specified material; statements such as "the dose rate in the irradiator is 1 kGy per hour" are therefore meaningless unless the nature of the absorbing material is also specified. By convention, water is generally taken as the reference material and dosimeters will often be calibrated to read absorbed dose to water even though they are made of some other material.

Photons

For low atomic number materials the predominant form of interaction with photons of energies between 200 keV and 20 MeV is a process of inelastic scattering in which part of the energy of the photon is transferred to an electron ejected from an atom. This process, known as **Compton scattering**, results in a lower energy photon, which can go on to further Compton events, and an energetic electron which transfers its energy to the medium in a series of interactions with valence electrons. Two other processes by which a photon can transfer energy are the **photo-electric effect** and **pair-production**, but these only become important at energies lower and higher, respectively, than are used in radiation processing and will not be considered here. The fraction of energy deposited by a photon as it traverses unit mass of material is expressed by a quantity known as the **mass energy absorption coefficient** (μ_{en}/ρ). If the energy fluence in the photon beam (in terms of energy per unit area) is

Dosimetry for Food Irradiation

represented by the symbol Ψ then

$$\text{Absorbed dose} = \Psi \bullet (\mu_{en} / \rho)$$
$$\mu_{en} / \rho = 0.0309 \text{ cm}^2\text{g}^{-1} \text{ for 1 MeV photons in water}$$

The mass energy absorption coefficient is dependent both on the energy of the photon and the nature of the absorbing material. This means that different materials placed in the same photon beam will, in general, absorb different amounts of energy. As an example of this the mass energy absorption coefficient of glass is about 10% lower than that of water for a typical degraded ^{60}Co spectrum; tables of mass energy absorption coefficients can be found in many text books on radiation dosimetry.

Photons in the energy ranges used for radiation processing show an approximately exponential decrease of absorbed dose with depth; the exact shape of the curves is dependent on the photon energy, the size of the source and the distance between the source and the target. Figure 1 shows a typical depth dose curve for ^{60}Co radiation in water.

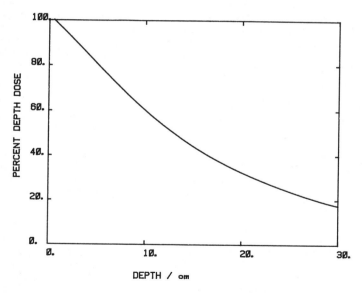

Figure 1 Central axis depth dose curve in water for ^{60}Co radiation, field size 20 x 20 cm, 80 cm source to surface distance

Under certain conditions the maximum absorbed dose is observed not at the surface of the target but at a small depth into it (a few mm in the case of

^{60}Co). This arises because more secondary electrons are lost from the surface of the material than are replaced by secondary electrons generated in the air between the source and target. As it is these secondary electrons that are responsible for most of the absorbed dose, the dose at the surface will be lower than in a continuous medium. At depths into the target greater than the range of secondary electrons an equilibrium will exist with electrons escaping from a given volume being replaced by an equal number entering the volume; such a condition is known as **charged particle equilibrium**. In practice the geometry of the source generally means that such a **build-up** of dose is rarely seen in radiation processing applications but similar effects due to loss of charge particle equilibrium can occur at boundaries between materials of different atomic number. These situations can be very complex and may involve an increase in dose at the surface rather than the decrease discussed above.

Electrons

High energy electrons lose energy in a continuous series of interactions with the valence electrons of the material through which they are passing. The rate at which energy is lost (in terms of energy per unit length) is given by a quantity known as the **stopping power** of the material; this quantity is often divided by the density of the material to give the **mass stopping power**. Energy lost by fast electrons can either be absorbed by the medium or radiated as bremsstrahlung radiation. The stopping power thus consists of two components, a collision component which describes interaction leading to absorption of energy close to the electron track and a radiative component which describes bremsstrahlung production. For the low atomic number materials generally involved in radiation processing the radiative component is essentially negligible.

Collision mass stopping power (S_{col}/ρ) is dependent on both electron energy and the nature of the absorber and thus different materials placed in the same electron beam will, in general, absorb different amounts of energy.

Figure 2 shows typical depth dose curves in water for broad electron beams of various energies. The position of the peak of absorbed dose is very sensitive to beam size and geometry and in many situations will appear much nearer the surface than shown here. Unlike photons, electrons have a finite range which, for water, amounts to approximately 5 mm per MeV.

Cavity Effects

Any interface between materials of different atomic numbers will result in a loss of charged particle equilibrium around the interface and hence a change in the absorbed dose. Unfortunately materials used for dosimetry are often of different atomic composition to the material in which it is required to measure dose, in other words, the dosimeter forms a **cavity** within the medium. The treatment of this problem is beyond the scope of this article but its effects should always be borne in mind especially when dealing with thin (<1 mm) dosimeters. Because most radiation processing involves low atomic number materials a simplification is often made by calibrating all dosimeters in terms of absorbed dose to water. Strictly, dosimeters calibrated in this way will only read absorbed

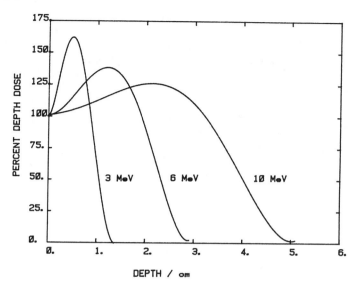

Figure 2 Dose vs Depth in water for broad electron beams of the indicated energies

dose to water when surrounded by sufficient water equivalent material to achieve charged particle equilibrium and problems still arise if a thin dosimeter is placed close to material of high atomic number.

Dose Distribution In Irradiated Products

The variation in absorbed dose with depth through an absorbing material has important consequences for the uniformity with which any product can be irradiated. In a typical ^{60}Co irradiator boxes of product will travel on a conveyer system and pass both sides of a rectangular source plaque in such a way that a given product box is effectively irradiated from two opposite sides. The dose distribution through such a box can, to a first approximation, be represented by the sum of two exponential depth dose curves obtained as the box passes either side of the source. This situation is shown schematically in Figure 3. Product at the edge of the box will have received the highest dose, D_{max}, whilst product in the centre will have received the least dose, D_{min}. The ratio of maximum to minimum dose within a box, D_{max}/D_{min}, is generally known as the overdose ratio and its value will depend on the geometry of the irradiator and the dimensions and weight of the product box. Typical values of overdose ratios are around 1.5 - 2, although values in excess of 3 are not unknown.

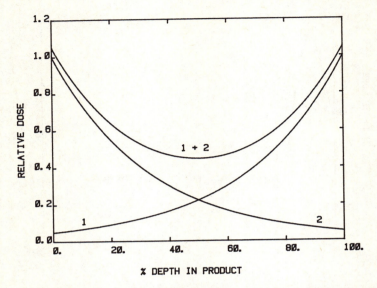

Figure 3 Schematic representation of dose distribution through a product box following two sided ^{60}Co irradiation

In addition to the maximum and minimum dose received by product, food technologists are often interested in the average dose received; indeed, the recommendation by the WHO/FAO/IAEA Joint Expert Committee on Food Irradiation was that food irradiated up to an "overall average" of 10 kGy should generally be released for human consumption as long as the maximum dose does not exceed 15 kGy. Whilst the concept of average dose clearly has attractions from the point of view of feeding studies, it is very difficult to measure in practice and requires the distribution of a large number of dosimeters throughout the product box. Because of this difficulty an approximation is often made for gamma irradiation and average dose taken as $(D_{max}+D_{min})/2$. This approach becomes increasingly inaccurate as the overdose ratio increases and will lead to average dose being over-estimated by around 20% at an overdose ratio of 3[1].

Dose variations in electron beam irradiation are much more difficult to predict and each type of product container has to be individually "dose mapped". The concept of overdose ratio is still valid although the relationship between D_{max}, D_{min} and $D_{average}$ is not straightforward.

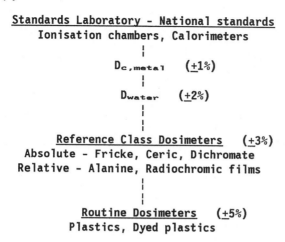

Figure 4   A typical national standardising chain

3. STANDARDS DISSEMINATION CHAIN

Radiation dosimeters can be classified according to their position in a standards hierarchy. Figure 4 shows a typical national standardising chain starting at the top with the national standards, usually primary standards, held at the national standards laboratory. These **primary standards** are either ionisation chambers or calorimeters and will measure absorbed dose to carbon or metal with an uncertainty of ±1% or less. A theoretical step is required to convert to the quantity normally used, absorbed dose to water, and this will typically add about another 1% to the overall uncertainty.

The next stage in the standards chain is a class of dosimeters known as **reference dosimeters**. These are chemical dosimetry systems whose response not only to radiation but also to other influencing factors such as temperature and dose rate is reproducible and well characterised. Reference dosimeters can be divided into two categories, **absolute** and **relative**. Absolute dosimeters, for example Fricke, ceric and dichromate solutions, are systems whose radiation chemical response is characteristic of a particular solution composition and this can be guaranteed, certainly to within a few percent, provided good chemical practice is followed. Relative dosimeters, for example alanine/ESR and radiochromic films, are systems which, although exhibiting high precision, have to be calibrated against a standard higher in the chain. Their radiation response cannot be predicted with sufficient accuracy purely on the basis of composition. The overall uncertainty associated with reference dosimeters is about ±3%.

Routine dosimeters are systems whose performance, particularly with respect to environmental influencing factors, is not as good as the reference systems, but whose ease of use and low cost makes them ideal for the day-to-day monitoring of radiation processes. These dosimeters are generally based on plastics and usually rely on radiation induced changes in optical absorption as the method of measurement. Uncertainties associated with routine dosimeters are at best around the ±5% level but environmental influence factors such as temperature and humidity can lead to large systematic errors if not taken properly into account (see later).

4. REFERENCE DOSIMETERS

Some of the more commonly used reference dosimeters are listed below with brief descriptions of their composition and method of use. Detailed operating procedures for individual systems can be found in the references cited.

Many national standards laboratories, and also the International Atomic Energy Agency (IAEA), offer services in which reference dosimeters are mailed to customers who irradiate them and then return them to the originating laboratory for measurement and dose evaluation. Such services provide a vital link in the standards dissemination chain and ensure that routine radiation processing dosimetry is **traceable** to national primary standards.

Fricke Dosimeter[2,3]

The Fricke dosimeter is the best known and most well characterised of all chemical dosimeters and is the only dosimeter, to date, for which the ICRU have recommended a radiation chemical response. Many institutions rely on Fricke as their absolute standard of absorbed dose although if this is done it is essential that estimates of dose are checked by some form of intercomparison with a recognised standards laboratory.

The standard Fricke dosimeter consists of an aerated solution of 1 mM ferrous ammonium sulphate in 0.4 M sulphuric acid. Fricke solution, like all liquid chemical dosimeters, is sensitive to trace levels of impurities, particularly organic contaminants, and great care has to be taken in the preparation of the dosimeter. Sodium chloride (1 mM) is often added to the solution to suppress the effect of impurities but this does reduce the maximum dose rate that can be used. The overall reaction occurring in the Fricke dosimeter is an oxidation of ferrous ion (Fe^{2+}) to ferric ion (Fe^{3+}); the concentration of the latter being measured by spectrophotometry at 304 nm.

In common with most aqueous chemical dosimeters, the mass fraction of the active solute is so low that the effect of direct radiation interaction on solute can be ignored. The initial interaction is between radiation and water which, in acid conditions, will result in the formation of hydrogen atoms (H), hydroxyl radicals (OH), hydrogen peroxide (H_2O_2) and molecular hydrogen (H_2). The chemistry of the Fricke dosimeter is unusual in that each hydrogen atom, normally regarded as a reducing agent, reacts to oxidise three ferrous ions. The detailed mechanism is complex but can be represented in a simplified form as:

$$H + O_2 \longrightarrow HO_2$$
$$HO_2 + Fe^{2+} + H^+ \longrightarrow Fe^{3+} + H_2O_2$$
$$Fe^{2+} + H_2O_2 + H^+ \longrightarrow Fe^{3+} + OH + H_2O$$
$$Fe^{2+} + OH \longrightarrow Fe^{3+} + OH^-$$

The reaction of OH radicals with ferrous ions is relatively slow and problems can arise in the presence of trace organic material as OH radicals will react rapidly to form organic radicals which, in turn, can go on to take part in chain reactions involving ferrous ions. Chloride ion is an extremely efficient scavenger of OH radicals thus blocking the reaction of OH radicals and organic material. The resultant chloride radicals are capable of oxidising ferrous ion but will not react to form organic radicals.

The Fricke dosimeter shows a linear response over the dose range 20 to 200 Gy, but can be used up to 400 Gy if account is taken of the non-linearity. The temperature coefficient during irradiation is +0.1% per degree around room temperature and solutions without sodium chloride are independent of dose rate up to about 2 MGy/s.

A variant of the dosimeter, known as the Super Fricke, uses 10 mM ferrous ion and is oxygen saturated. This overcomes some of the problems which occur in the normal Fricke at very high doserates and extends the usable range to about 30 MGy/s.

Ceric / Cerous Dosimeters[3,4]

Dosimeters based on the reduction of ceric ion (Ce^{4+}) to cerous ion (Ce^{3+}) are used extensively for determining absorbed dose in the kilogray dose range. The solutions are made up in 0.4 M sulphuric acid and a range of ceric ion concentrations are used depending on the dose to be measured (eg 10 mM for the 10 - 40 kGy region). Cerous ion (10 mM) is often added to the solution to reduce the effect of organic impurities and the solution is used air saturated. The temperature coefficient during irradiation is approximately -0.2% per degree, ie the response decreases as the temperature increases. The dosimeter is independent of dose rate up to about 1 MGy/s.

As with the Fricke dosimeter the effect of radiation is to create H atoms, OH radicals and H_2O_2 from radiolysis of the water and it is these species which go on to react with cerium ions. Unlike Fricke, the mechanism involves both oxidation and reduction, the main reactions can be represented as:

$$Ce^{4+} + H \longrightarrow Ce^{3+} + H^+$$

$$2Ce^{4+} + H_2O_2 + 2H^+ \longrightarrow 2Ce^{3+} + 2H_2O$$
$$Ce^{3+} + OH \longrightarrow Ce^{4+} + OH^-$$

The overall amount of ceric ion reduced is the difference between the radiolytic yields of reducing species (H atoms and H_2O_2) and oxidising species (OH radicals). The addition of 10 mM cerous ion, mentioned above, reduces the probability of OH radicals reacting with organic impurities.

Readout can either be by spectrophotometry or by a potentiometric method. The ceric ion exhibits an optical absorbance peak at 320 nm which enables the reduction in ceric ion concentration to be measured, although dilution of the solution by a factor of 100 is necessary to obtain a reasonable optical density. An easier, but less accurate, potentiometric method involves placing irradiated and unirradiated solution in a concentration cell separated by a thin membrane. The difference in the Ce^{4+}/Ce^{3+} ratio in the two solutions results in an EMF generated between electrodes placed in each solution.

Dichromate Dosimeter[5,6,7]

Dosimeters based on the reduction of Cr(VI) to Cr(III) in aqueous solution can be used in the dose range 2 to 50 kGy. Initial dichromate ion concentrations of 0.5 mM and 2.5 mM are required to cover the ranges 2 to 10 kGy and 10 to 50 kGy respectively. The solutions are made up in 0.1 M perchloric acid and contain 1 mM silver ion to prevent side reactions which result in increased response at high dose rates. The temperature coefficient during irradiation is around -0.2% per degree and the dosimeter is independent of dose rate in pulsed beams up to 100 MGy/s.

The mechanism of the dichromate dosimeter is very similar to that of the ceric dosimeter in that H atoms and H_2O_2 will reduce Cr(VI) to Cr(III) whilst OH radicals will cause the reverse. Again an OH radical scavenger, Ag^+ ions in the case of the dichromate dosimeter, is used to prevent unwanted side reactions.

The decrease in concentration of Cr(VI) with increasing radiation dose can be measured by spectrophotometry without dilution of the solution. Two wavelengths are used, 350 nm for an initial concentration of 0.5 mM dichromate and 440 nm for an initial concentration of 2.5 mM. The response of the dosimeter deviates only slightly from linearity up to 95% conversion.

Alanine / ESR Dosimeter[8]

The alanine dosimeter is the best known of a growing number of dosimeters that rely on Electron Spin Resonance (ESR) methods to determine radical concentrations in irradiated solids. The unpaired electron associated with a free radical can exist in two quantised spin states. Under normal conditions the energy of these two states is identical, but if the sample is placed in a strong magnetic field interactions between the field and the electron spin result in one spin state having a higher energy than the other. If a photon of the right energy is applied to the sample it will be absorbed and cause a transition from one spin

Dosimetry for Food Irradiation 119

state to the other in exactly the same way as occurs in any other form of spectroscopy. Commercial ESR spectrometers generally work at magnetic field strengths of a few hundred milli-tesla which leads to an energy difference corresponding to photons in the microwave region.

Although free radicals are produced in most substances on irradiation they are not generally stable and usually undergo rapid reaction. One exception to this is the amino acid alanine (CH_3CHNH_2COOH) in which trapped free radicals are stable over many years provided steps are taken to exclude water. This stability has led to the development of several types of dosimeter in which the intensity of ESR absorption from microcrystalline alanine is used to determine radiation dose. The alanine may be used as a pure powder or it may be pressed into pellets or incorporated into polymer matrices depending on the application. The usable dose range of the system is very wide (10 Gy to 100 kGy) and the temperature coefficient during irradiation around 0.2% per degree.

Radiochromic Dye Films[9]

Radiochromic dye film is a generic name for a whole family of dosimeters based on dye precursor molecules trapped in thin polar polymeric films. The films are generally around 40 to 60 µm in thickness and operate in the 1 to 100 kGy dose range. Measurement is by optical absorption in the region around 600 nm (the exact wavelength depends on the type of dye used). The dye precursors are triphenylmethane cyanide (Ph_3C-CN) derivatives with a variety of functional groups as substituents on the benzene rings. On irradiation the cyanide group is lost to leave a highly conjugated carbonium ion. The most commonly used polymer host is nylon, but polyvinyl butyral and polyhalostyrenes have also been used.

Although the main use of dye films is in determining dose gradients over very small distances, for example at an interface between dissimilar materials, their stability after irradiation makes them suitable for use as reference dosimeters in certain situations. Dye films used for reference dosimetry are usually packaged between two pieces of plastic (PMMA or polystyrene) several millimetres in thickness to provide charged particle equilibrium and must be sealed in a water-proof sachet to ensure no changes in water content. The temperature coefficient during irradiation has been found to vary between batches of material and should be determined for the batch in use.

5. ROUTINE DOSIMETERS

Although many different dosimetry systems are in use for the routine monitoring of radiation processing, it is possible to divide them into several main types. In the following sections each of these main types will be described in general terms and examples given of commercially available systems; the list is not intended to be exhaustive and no conclusions should be drawn from the inclusion or not of a commercial dosimeter.

Dosimeters Based On Undyed Plastics[10]

Radiation degradation of polymers often results in the generation of unsaturated molecules which absorb in the UV region of the spectrum. Polymethylmethacrylate (PMMA, Perspex), in particular, exhibits changes in optical absorbance around 300 nm and several commercial dosimetry systems are available (Perspex HX, Radix) for use in the 5 kGy to 50 kGy dose range. The dosimeters are generally cut into strips 10 mm wide and between 1 and 3 mm thick which can easily be mounted in commercial UV spectrophotometers. The strips are normally supplied sealed in aluminium foil sachets for protection. The method of manufacture of the dosimeters varies and some show large thickness variations which require the thickness of each dosimeter to be measured and a correction applied.

Another polymer which finds use as a dosimeter is cellulose tri-actetate (CTA) which is available as a thin film 125 μm thick, 8 mm wide and in lengths up to 100 m. The film is particularly useful in electron beam applications where dose profiles can easily be determined by scanning along a length of film in a modified spectrophotometer.

Polyvinyl chloride (PVC) film has been used as dosimeter in electron beam processing for many years but experience has shown the system to have many drawbacks and it cannot be recommended for anything but rough dose estimation. Readout is by spectrophotometry at 395 nm but the film has to be "developed" by heating for 1 hour at 50°C, or 10 minutes at 80°C before readout.

Dosimeters Based On Dyed Plastics[10]

Dyed plastics probably form the single largest group of routine dosimeters and within this group dyed PMMA systems are probably more common than any other. All systems rely on reaction of dye, or dye precursor, molecules with species formed from irradiation of the polymer matrix. The products formed, often free radicals, have optical absorption bands in the visible part of the spectrum and thus the spectrophotometry is generally easier than at the UV wavelengths encountered with undyed systems.

Several systems (Red 4034, Amber 3042, GAMMACHROME YR and Gammex) are supplied in the the form of 10 mm wide strips of 2 to 3 mm thickness which can easily be fitted into commercial spectrophotometers. These are supplied sealed in foil sachets both for protection and to ensure the water content of the polymer does not change. The dose range covered by a system is determined by the type of dye used and dosimeters of this form are now available to cover the range from 100 Gy to 50 kGy. Some systems are moulded (eg Gammex) and show very little variation in thickness whilst others are cut from cast sheets (eg Red 4034) and show considerable variation which must be corrected for by thickness measurements on each dosimeter. The response of many systems is sensitive to the water content of the polymer; such systems are conditioned to a specific water content by the manufacturer and sealed into foil sachets to maintain these conditions.

Dosimetry for Food Irradiation

Radiochromic dye films, discussed above in *Reference Dosimeters* are also used as routine dosimeters, especially in electron beam applications. They are available commercially both in the form of 1 cm squares and also in larger sheets or rolls. These dosimeters are sensitive to water content and, as they are not supplied in sealed sachets, it is up to the user to ensure that they are not subject to wide variations in humidity.

A form of radiochromic dosimeter that warrants special mention is the Optichromic Dosimeter. This dosimeter consists of a dye precursor incorporated into a gel which is sealed into a plastic tube. The tube is closed at either end by a small glass bead and the whole assembly acts as an optical wave guide. Readout can either be in a modified spectrophotometer or in a custom designed reader. Optichromic dosimeters are available in tubes 3 mm diameter and 5 cm long covering the dose range 50 Gy to 50 kGy.

6. FACTORS AFFECTING RESPONSE OF ROUTINE DOSIMETERS

Routine dosimeters are, by definition, subject to a number of influence factors which affect their response to radiation. The most important of these are listed below together with some recommendations for minimising their effects.

Temperature

Temperature is probably the most significant external factor influencing the response of dosimeters. No dosimeter has yet been devised which does not show some dependence on temperature. Reference dosimeters usually have two well characterised temperature coefficients - during irradiation and measurement respectively - which have to be taken into account. The measurement temperature can, of course, be controlled and as long as the irradiation temperature is known to within 5°C, errors of not more than about 1% should result.

The situation with routine dosimeters is much more complex as many are sensitive to the temperature **after** irradiation as well as the temperature **during** irradiation. In addition, the important parameter after irradiation is not merely the temperature but the length of time at a given temperature and thus it is impossible to devise any simple correction factor for the dosimeter. The only solution is to calibrate the dosimeters under conditions of temperature, and time, as close as possible to the conditions under which they will be used. As an example of the magnitude of the errors which can occur, Red 4034 Perspex irradiated to 25 kGy over a period of 6 hours at 50°C will over-estimate dose by 10% if read against a calibration curve prepared at 25°C.

Humidity

The response of all polymer dosimeters is influenced by the water content of the polymer and, hence, by the humidity at which it is stored. Many dosimeters are preconditioned to a specific water content by the manufacturer and then sealed in foil sachets to maintain those conditions. If the sachet is damaged then the water content of the dosimeter is likely to change and with it the response of the dosimeter. Thin film dosimeters pose particularly severe problems as they re-equilibrate with atmospheric humidity very rapidly. Because of this thin films should be stored in an humidity controlled environment until immediately prior to use.

Fading

Chemical reactions continue in most routine dosimeters for many days after irradiation and this accounts for the post-irradiation temperature effects discussed above. Even under stable temperature conditions changes in optical density after irradiation mean that it is often necessary to prepare a series of calibration curves for various readout times. As post irradiation changes are most rapid immediately after irradiation it is often advantageous to wait 12 to 24 hours before reading a dosimeter.

Dose Rate

Dose rates encountered in radiation processing can vary over many orders of magnitude with equivalent doses being delivered over 12 hours or so in a ^{60}Co plant or in a matter of a few seconds in an electron accelerator. Most dosimetry systems show some change in response as dose rates increase above a certain level but can be considered essentially independent of dose rate below that level. In general, very little dose rate dependence is seen at γ- or X-ray dose rates but pronounced effects are seen at electron accelerator dose rates. This means that dosimeters used in electron beams must be calibrated in electron beams.

BIBLIOGRAPHY

F H Attix,
'Introduction to Radiological Physics and Radiation Dosimetry'
Wiley-Intersciences, New York, 1986

W L McLaughlin, A W Boyd, K H Chadwick, J C McDonald and A Miller,
'Dosimetry for Radiation Processing'
Taylor and Francis, London, 1989

'Manual on Food Irradiation Dosimetry'
Technical Reports Series, No. 178
International Atomic Energy Agency, Vienna, 1977

REFERENCES

1. K H Chadwick and W F Oosterheert,
   Int. J. Appl. Radiat. Isot., 1986, 37, 47

2. 'Method for Using Fricke Dosimetry to Measure Absorbed Dose in Water'
   American Society for Testing and Materials, Standard E 1026

3. R W Matthews,
   Int. J. Appl. Radiat. Isot., 1982, 33, 1159

4. 'Method for Using the Ceric-Cerous Sulfate Dosimeter to Measure Absorbed Dose in Water'
   American Society for Testing and Materials, Standard E 1205

5. P H G Sharpe, J H Barrett and A M Berkley,
   Int. J. Appl. Radiat. Isot., 1985, 36, 647

6. P H G Sharpe and K Sehested,
   Radiat. Phys. Chem., 1989, 34, 763

7. P H G Sharpe, A Miller and E Bjergbakke,
   Radiat. Phys. Chem., 1990, 35, 757

8. D F Regulla and U Deffner,
   Int. J. Appl. Radiat. Isot., 1982, 33, 1101

9. W L McLaughlin, A Miller, S Fidan, K Pejterson and W Batsberg Pederson,
   Radiat. Phys. Chem., 1977, 10, 119

10. J H Barrett,
    Int. J. Appl. Radiat. Isot., 1982, 33, 1177

The Effects of Ionising Radiation on Additives Present in Food-contact Polymers

D. W. Allen, A. Crowson, D. A. Leathard and C. Smith

DEPARTMENT OF CHEMISTRY, SHEFFIELD CITY POLYTECHNIC, SHEFFIELD S1 1WB, UK

1 INTRODUCTION

There is considerable current interest in the possible use of ionising radiation as a means of food preservation. Irradiation can be used to kill or reduce the numbers of pathogenic or spoilage organisms in food and to control infestation in stored products. In its recent report, the Advisory Committee on Irradiated and Novel Foods[1] (ACINF) has advised that the irradiation of food up to an overall average dose of 10 kGy presents no toxicological hazard and introduces no special nutritional or microbiological problems. The UK Government has accepted the recommendations of the ACINF report, and the introduction of food irradiation in the UK is likely in the near future.

Whereas many types of irradiated foods have been studied in depth, and the effects of irradiation on food contact plastics have been investigated in some considerable detail,[2] there is much less information on the effects of irradiation on the many additives present in food-contact plastics, although it has been established that changes do occur in the migration behaviour of such additives.[3] It is possible that toxic substances might be formed within the polymer, and could subsequently migrate into the foodstuff, thereby presenting a hazard to the consumer. In studying this area, our aims have been to study the effects of irradiation on a range of polymer-stabilising additives in terms of (i) the influence of irradiation dose on the extent of destruction of the additive, (ii) the identification of extractable degradation products derived from the additives, (iii) the influence of irradiation dose on the extent to which the additives and their mobile degradation products are able to migrate into food-simulant media, and also (iv) to compare the effects of gamma- and electron-beam irradiation on the above.

2 RESULTS AND DISCUSSION

In early work, we showed that organotin stabilisers of the type Bu_2SnX_2 ($X = SCH_2CO_2C_8H_{17}$ or $O_2CCH=CHCO_2C_8H_{17}$) present in poly(vinyl chloride) (PVC) and subjected to varying doses of gamma irradiation in the range 1-200 kGy (0.1-20 Mrad) suffer degradation with dealkylation to form monobutyltin trichloride and tin(IV) chloride (Tables 1 and 2)*, which have been characterised by a subsequent alkylation procedure followed by gas chromatographic analysis. The extent of degradation of the stabilisers on prolonged gamma irradiation is much more severe than during thermal degradation leading to comparable blackening of the polymer.[4-7]

Table 1 Analysis of Gamma-Irradiated PVC Samples containing dibutyltin bis(iso-octylthioglycollate) (1.2% w/w) [$Bu_2Sn(IOTG)_2$]

Exposure /kGy	Relative proportion of degradation products[a] (%)		
	Bu_2SnX_2	$BuSnX_3$	$SnCl_4$
0	92	6	2
5	87	10	3
10	88	7	5
15	75	15	10
20	72	17	11
25	70	16	14
50	52	14	34
100	35	18	47
200	15	17	68

[a] $X = SCH_2CO_2C_8H_{17}$ or Cl

*Reproduced from "Applied Organometallic Chemistry" with permission of The Longman Group UK.

Table 2 Analysis of Gamma-Irradiated PVC Samples (Milled) containing Dibutyltin bis(iso-octylmaleate) (2% w/w) [$Bu_2Sn(IOM)_2$]

Exposure /kGy	Relative proportion of degradation products[a] (%)		
	Bu_2SnX_2	$BuSnX_3$	$SnCl_4$
0	97	3	1
5	91	5	4
10	90	7	3
15	92	4	4
20	92	4	4
25	89	6	5
50	68	16	16
100	59	22	19
200	41	17	42

[a] $X = O_2CCH=CHCO_2C_8H_{17}$ or Cl

In more recent work, we have studied the fate of hindered phenol and arylphosphite antioxidants (e.g. Irganox 1076, 1010, and 1330, and Irgafos 168) (Scheme 1) present in a range of polymers, including PVC, polyethylene and polypropylene which have been subjected to varying doses of gamma- and electron beam-irradiation. Such antioxidants are present not only to stabilise the polymer during initial processing and fabrication, but also during subsequent service, their main role being the removal of alkoxy and alkylperoxo radicals which would otherwise lead to degradation of the polymer. These compounds would be expected to have an important role to play in the suppression of the ambient-temperature oxidation of polyolefins following irradiation.[8-11]

Our hplc analytical results revealed the progressive destruction of the antioxidants on gamma-irradiation, the rate of loss depending on the nature of both antioxidant and base polymers (Tables 3 and 4). In most cases, a significant (but not drastic) loss of antioxidant (ca. 30%) occurs on exposure to a gamma irradiation dose of 10 kGy (1 Mrad), the maximum dose likely to be permitted should food irradiation gain general approval in the UK. Reanalysis of the polymers after a six-month interval has revealed little subsequent post-irradiation degradation of the antioxidants. The extent of degradation of the antioxidants after a 25 kGy dose is correspondingly greater.[12]

Table 3 Effects of Gamma-irradiation on Phenolic Antioxidants Present in poly(vinyl chloride), polyethylene and polypropylene

Irradiation Dose/kGy	PVC		Polyethylene		Polypropylene	
	Irganox 1076[a]	Irganox 1010[b]	Irganox 1076[a]	Irganox 1010[b]	Irganox 1076[a]	Irganox 1010[b]
0	0.44	0.62	0.36	0.16	0.34	0.08
1	0.29	0.46	0.28	0.12	0.37	0.07
5	0.27	0.46	0.23	0.10	0.38	0.05
10	0.12	0.43	0.22	0.09	0.38	0.04
20	0.17	0.37	0.20	0.07	0.36	0.02
25	0.18	0.31	0.14	0.07	0.35	0.02
35	0.19	0.30	0.14	0.05	0.30	0.01
50	0.15	0.24	0.11	0.04	0.30	0.01

[a] ±0.03%. [b] ±0.01%.

Table 4 Effects of Gamma-irradiation on Irgafos 168 Present in Polypropylene

Irradiation Dose/kGy	Irgafos 168 as sole antioxidant %	Combination of Irgafos 168 + Irganox 1010 %	
	Irgafos 168[a]	Irgafos 168[a]	Irganox 1010[b]
0	0.067	0.069	0.08
1	0.035	0.010	0.05
5	0.009	-	0.04
10	0.004	-	0.03
20	-	-	0.02
25	-	-	0.01

[a] ±0.003. [b] ±0.01.

In the case of the hindered phenol antioxidants, we did not initially detect significant quantities of degradation products which could be extracted from the irradiated polymer and it was suspected that such products were becoming covalently bound to the polymer as a result of radical coupling processes. It has recently been shown that gamma-irradiation of hindered phenols in benzene solution gives rise to phenylated derivatives resulting from coupling of radicals derived from the antioxidants with phenyl radicals derived from the solvent.[13] Gamma-irradiation of polyolefins

(1) Irganox 1076: HO–[3,5-di-t-Bu-C$_6$H$_2$]–CH$_2$CH$_2$CO$_2$C$_{18}$H$_{37}$

(2) Irganox 1010: [HO–(3,5-di-t-Bu-C$_6$H$_2$)–CH$_2$CH$_2$CO$_2$CH$_2$]$_4$C

(3) Irganox 1330

(4) Irgafos 168

(Scheme 1)

is known to give rise to macroalkyl radicals, and hence the trapping of antioxidant degradation products is therefore probable. If this is so, then concerns over the migration of potentially toxic degradation products are much reduced.

Experiments involving a ^{14}C-labelled antioxidant in irradiated polyolefins have given some support to this suggestion. A sample of Irganox 1076 labelled with ^{14}C at the benzylic carbon of the arylpropionic acid moiety was diluted with unlabelled Irganox 1076, and the diluted, labelled antioxidant incorporated into both polypropylene and high density polyethylene at ca. 0.2% by weight. Following irradiation, the total extractable ^{14}C-activity was measured, and found to decline progressively as the irradiation dose increased for both polymer systems. At the same time, the residual ^{14}C-activity of the exhaustively-extracted polymers was found to increase with increasing dose, indicating the probable binding of at least 20% of the total available ^{14}C-activity to the polymer after a 50 kGy exposure. In order to explore the possibility of physical entrapment of antioxidant and degradation products in the possibly cross-linked, irradiated polymer, which could, in itself, result in the observed reduction in extractable radioactivity, a 50 kGy-irradiated sample of the ^{14}C-labelled polypropylene was dissolved in hot tetralin, followed by precipitation of the polymer by the addition of hexane. It was considered that any unbound antioxidant (and degradation products) would be removed in the tetralin fraction, whereas the bound substances would be precipitated with the polymer. The precipitated polymer fraction was repeatedly washed with hexane to remove any residual unbound activity, and its ^{14}C-activity then assayed. The results of this procedure indicated that at least 12.4% of the total ^{14}C-activity had become polymer-bound as a result of the irradiation process. The true figure is likely to be significantly higher (as indicated by the above residual activity figure), as a result of losses of labelled low molecular weight oligomeric fractions into the supernatant liquid in the above dissolution-precipitation procedure. This investigation could not be repeated using high density polyethylene because, following irradiation, this material became impossible to dissolve in hot tetralin, presumably as a result of cross-linking. Thus, although evidence of binding of antioxidant-degradation products to the irradiated polymer has been obtained, the extent of the binding is lower than might have been expected in view of our initial failure to detect extractable degradation products by hplc techniques. In subsequent studies, we have now refined our chromatographic procedures, and have observed a number of extractable degradation products arising from the above range of hindered phenol antioxidants. Thus, in the case of Irganox 1076, one significant degradation product has been detected, whereas for Irganox 1010, three major and several minor degradation products have been detected. Similarly, for Irganox 1330, several degradation products are apparent. The challenge now is to isolate and characterise these substances.

In the case of the arylphosphite stabiliser, Irgafos 168, we have shown that drastic reductions in the level of the antioxidant occur during gamma irradiation, to such an extent that little remains after a dose of 10 kGy (Table 4). In addition, we have detected the triarylphosphate oxidation product of Irgafos 168 in the extracts of the irradiated polymers by both hplc and ^{31}P nmr techniques. As the irradiation dose increases, it would appear that the phosphorus-containing degradation products are also becoming covalently bound to the polymer. However, there is also evidence of the formation of a range of extractable degradation products in increasing amounts as the dose increases.

We have also studied the effects of gamma irradiation on the extent of migration of hindered phenol antioxidants into fatty food simulants.[14] It is generally considered that the migration of polymer additives into foodstuffs is a diffusion problem, and depends on many variables.[15] As the determination of migrated polymer additives in heterogeneous foodstuffs is a difficult and time-consuming task, it has become common practice to study the migration of polymer additives into a series of homogeneous liquids (food simulant media), under standard conditions, e.g. a polymer surface area of 2 dm^2 exposed to 100 cm^3 simulant over a period of 10 days at 40 °C. A range of oily, alcoholic and aqueous simulants specified by EEC Directive,[16] has been widely used in such studies.

The specific systems studied by us to date have involved the migration of Irganox 1076 and Irganox 1010 present in polyolefins into the synthetic triglyceride fatty food simulant HB307 (10 days at 40 °C) and for comparison, into iso-octane (2 days at 20 °C). The latter has been proposed[17] as a fatty food simulant which provides a convenient alternative to fats such as HB307 and olive oil in that determination of migration into iso-octane is significantly easier, thereby providing a fast, cheap and simple predictive method for assessment of migration into fatty media. However, criticisms have been made of the use of iso-octane for this purpose.[18,19]

Samples of polypropylene in sheet form, containing Irganox 1076 and 1010, respectively at ca. 0.2% by weight, were prepared by conventional hot-milling and compression moulding techniques. These were used in studies of the extent of migration into iso-octane at 20 °C over 2 days, the antioxidants being determined by hplc techniques. For studies of migration into the synthetic fat HB307 at 40 °C over a period of 10 days, a ^{14}C-labelled Irganox 1076 antioxidant was similarly incorporated into both polypropylene and high density polyethylene (HDPE). The extent of migration was assayed by conventional liquid scintillation techniques. In separate experiments, it was demonstrated by TLC techniques that the ^{14}C-activity migrating into the simulant was due predominantly to unchanged ^{14}C-labelled Irganox 1076, although small amounts of another, as yet unidentified substance could also be detected. We therefore feel justified in assuming that the ^{14}C-activity of the simulant reflects the degree of migration of the antioxidant. Prior to the migration experiment, the polymer samples were subjected to varying doses of gamma-irradiation from a Cobalt-60 source. The results of the migration experiments are presented (Table 5) as specific migration values (mg antioxidant migrated per dm^2 contact area).

It is clear that the extent of migration into the iso-octane and HB307 decreases steadily as irradiation progresses, consistent with the reduction in the amount of extractable antioxidant revealed in our earlier study.[12] These results are also consistent with an earlier report[20] of the effects of a 25 kGy gamma dose on the migration of Irganox 1076 into HB307. It is of interest that the extent of migration of Irganox 1076 into iso-octane is significantly greater than that of Irganox 1010 reflecting the greater lipophilicity of the former. Furthermore, while the extent of migration of Irganox 1076 into iso-octane is greater than into HB307 under the stated conditions, the results are nevertheless comparable in magnitude, thus providing some justification for the use of iso-octane as a convenient indicator simulant for migration into fatty foods. The key conclusion from our study, however, is that gamma irradiation leads to a decrease in the degree to which hindered phenol antioxidants migrate from polyolefins into fatty media.

Table 5   Effects of Gamma Irradiation on Migration of Antioxidants into Fatty Food Simulants

Irradiation Dose/kGy	Iso-octane		HB307 ^{14}C-Irganox 1076	
	Irganox 1076 mg dm^{-2} polypropylene	Irganox 1010 mg dm^{-2} polypropylene	mg dm^{-2} polypropylene	mg dm^{-2} HDPE
0	2.6	0.8	1.0	1.3
10	2.1	0.3	0.7	1.0
25	1.3	<0.2	0.5	0.7
50	0.4	<0.2	0.2	0.3

Electron-beam irradiation offers an alternative approach for the radiation treatment of pre-packaged foods as an on-line process, provided that due consideration is given to the energy of the incident radiation and to the thickness of the package. Electron-beam irradiation is carried out with high energy electron accelerators which allow the delivery of high irradiation doses over a very short period of time. Thus, whereas a typical ^{60}Co gamma source can deliver a dose rate of ca. 12 kGy per hour, an electron-beam facility can deliver several tens of kGy per second. This significant difference in dose rate might result, therefore, in changes in the chemical processes taking place in the stabilised irradiated polymer, and several groups have reported experiments which support this view. Azuma et al.[21] have shown that electron-beam irradiation is more effective than gamma irradiation in minimising the formation of volatile degradation products derived from oxidative degradation of polyethylene. Lox et al.[22] have discovered significant differences in the effects which gamma- and electron-beam irradiation have on the extent of migration of organotin compounds from PVC into water.

Hence, it was of interest to establish whether the two irradiation processes had similar effects on the extent of destruction of antioxidants present in food contact polymers. Preliminary details have been reported of a comparison of the effects of gamma- and electron-beam irradiation on the extent of destruction of Irganox 1076, Irganox 1010 and Irgafos 168 present in polypropylene and low density polyethylene (LDPE) samples.[23]

Samples of the polyolefins stabilised as above were prepared by sintering to produce small pellets which were then subjected to progressive doses of irradiation in air from a ^{60}Co gamma source or a 4.5 MeV Dynamitron Continuous DC electron-beam facility. Following irradiation, the antioxidants were extracted from the irradiated polymers by heating under reflux in chloroform. The levels of antioxidants present in the resulting extracts were determined by hplc techniques using appropriate internal standards (either Irganox 1076 or Irganox 1010, depending on the analyte). The results are presented in Table 6.

It is clear that there is a broad similarity in the effects of the two types of radiation on the extent to which the above antioxidants are consumed, in the various polymer samples. With the exception of Irgafos 168, which is destroyed rapidly on irradiation, it is significant that, irrespective of the nature of the radiation employed, an appreciable proportion of the original antioxidant (ca. 50%) remains unchanged after an irradiation dose of 10 kGy (1 Mrad), the maximum irradiation level likely to be permitted. In addition, a comparison of the chromatograms obtained from the extracts of both gamma- and electron beam-irradiated polymers also reveals a very similar pattern of degradation products arising from a given antioxidant.

A preliminary study of the extent of migration of Irganox 1076 into iso-octane from a polypropylene-polyethylene copolymer subjected to electron-beam irradiation, together with corresponding data for the gamma-irradiated material, also reveals a broadly similar situation (Table 7). It is of interest that these results for the copolymer are noticeably higher than those for polypropylene presented earlier (Table 5).

Table 6 Comparison of electron-beam and gamma-irradiation results (percentage of initial antioxidant remaining)

Polyolefin (% w/w of antioxidant incorporated)	Dose /kGy Type of irradiation	0	1	5	10	25	50
Polypropylene +0.25% Irganox 1076	Electron-beam Gamma	100 100	85 NA	79 NA	68 74	53 60	44 57
Polypropylene +0.25% Irganox 1010	Electron-beam Gamma	100 100	100 NA	72 NA	59 48	35 29	22 11
Polypropylene +0.10% Irganox 1010	Electron-beam Gamma	100 100	75 88	52 63	35 50	19 25	11 13
Copolymer* +0.25% Irganox 1076	Electron-beam Gamma	100 100	100 NA	94 NA	86 86	76 63	58 55
Copolymer* +0.25% Irganox 1010	Electron-beam Gamma	100 100	89 NA	71 NA	70 53	40 28	38 11
Polypropylene +0.10% Irgafos 168	Electron-beam Gamma	100 100	38 52	<16 13	<16 <13	ND ND	ND ND
LDPE⁺ +0.20% Irganox 1076	Electron-beam Gamma	100 100	97 100	88 88	79 81	59 58	32 35

NA = No analysis carried out; ND = not detectable; *the copolymer consists of approximately 5% polyethylene in a polypropylene matrix. ⁺LDPE = low density polyethylene.

In the course of the above studies of the fate of antioxidants present in food-contact polymers subjected to varying irradiation doses, we have developed hplc procedures for individual antioxidants, e.g. Irganox 1076, which have involved the use of one of the related antioxidants e.g. Irganox 1010, as an internal standard. However, for the determination of Irgafos

Table 7 Comparison of the Effects of Gamma- and Electron-Beam Irradiation on Migration of Irganox 1076 from a Polypropylene-Polyethylene Copolymer into Iso-octane (2 days at 20 °C)

Dose /kGy	Electron-Beam Irradiation /mg dm^{-2}	Gamma Irradiation /mg dm^{-2}
0	3.7	5.1
10	2.8	4.3
25	2.2	2.7
50	1.6	1.3

168 in the presence of Irganox 1010 in synergistically-stabilised systems, it was necessary to seek an alternative internal standard, and the triazine, Irganox 565, 2,4-bis(n-octylthio)-6-(4-hydroxy-3,5-di-t-butylphenylamino)-1,3,5-triazine (5) was selected because of its favourable retention behaviour relative to the two analytes under the conditions employed. In the analytical procedure, samples of polymer were extracted under reflux in chloroform, which had been spiked with Irganox 565 as internal standard. In the analysis of unirradiated, thermally processed polypropylene, Irganox 565 functioned as a perfectly acceptable internal standard. However, when applied to samples of gamma-irradiated polypropylene and low density polyethylene (LDPE), it was found that the internal standard was progressively destroyed during the reflux period.[24]

In the case of polypropylene which had received a dose of 1 kGy, the Irganox 565 was completely destroyed. With gamma-irradiated LDPE, the rate of destruction was slower but significantly dose-related, complete destruction only being observed after a dose of 50 kGy. In contrast, little degradation was observed in the presence of irradiated high density polyethylene. Investigation of the behaviour of Irganox 565 in the presence of gamma-irradiated-polypropylene or -LDPE, containing no other antioxidants, revealed that it is converted into the iminoquinone (6), identified by comparison with the authentic material prepared by oxidation of the triazine (5) with manganese dioxide in chloroform at room temperature.

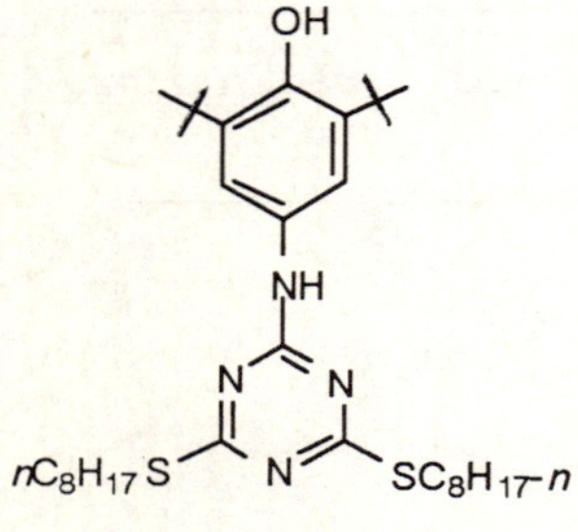

(5)

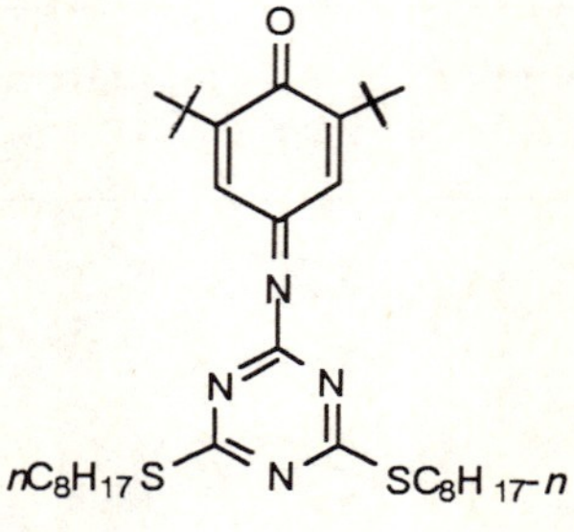

(6)

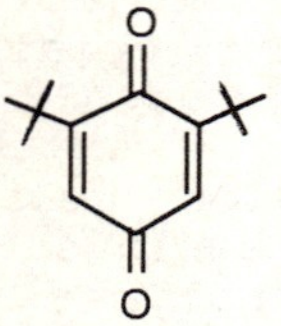

(7)

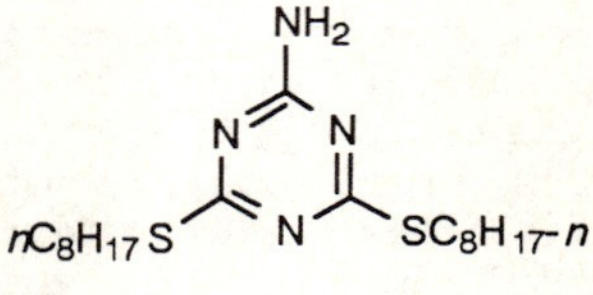

(8)

It is likely that the iminoquinone (6) is formed from the reactions of Irganox 565 with hydroperoxo groups formed at tertiary carbon sites in the irradiated polymers. Although hydroperoxides are formed during the thermal processing of polymers, it is clear that a much greater number of such groups are formed during gamma-irradiation in air. As pointed out above, little degradation of Irganox 565 occurs in the presence of an unirradiated (but thermally processed) polyolefin. The significantly smaller number of easily peroxidisable sites in irradiated HDPE doubtless accounts for the much reduced rate of destruction of the triazine (5). We have carried out model studies of the reaction of compound (5) with t-butyl hydroperoxide in chloroform solution, showing that the iminoquinone (6) is readily formed, together with the quinone (7) and the aminotriazine (8), arising from the hydrolysis of the iminoquinone (6) by traces of water in the peroxide reagent.

A very similar state of affairs applies to the decomposition of Irganox 565 by electron beam-irradiated polymers, the extent of decomposition again increasing from HDPE to LDPE to polypropylene. In all cases, the effect is dose-related.

In view of these findings, the possibility of a similar destruction of other internal standards used in the determination of antioxidants in irradiated polymers has also been investigated. It is found that although there is a small degree of decomposition of both Irganox 1076 and Irganox 1010 when used as internal standards, particularly in the presence of polypropylene subjected to large doses of irradiation (>25 kGy), the extent of this decomposition does not induce significant errors in the determination of the antioxidants discussed above. However, this study has indicated that Irgafos 168 is also rapidly destroyed when heated in solution in the presence of irradiated polymers. Similarly, Irganox 1330 is also subject to destruction at a similar rate to Irganox 565. Hence, in the determination of antioxidants in irradiated polymers, it is essential to select an internal standard which is relatively stable under the conditions employed. An additional problem is the related destruction of the analyte during solvent extraction from the irradiated polymer. An ideal internal standard for a given antioxidant analyte would therefore be a substance which suffers degradation at about the same rate as the analyte. Clearly, there are many challenges to the analytical chemist in this field!

As this Conference has shown, there is considerable interest in the development of tests for irradiated food. The above, "unexpected" chemistry may indicate an alternative approach in that it may be possible to devise reagents which would reveal whether a food-contact plastic has been irradiated.

ACKNOWLEDGEMENTS

This work has been supported by funds provided by the UK Ministry of Agriculture, Fisheries and Food. The results are the property of MAFF and are Crown Copyright.

The authors are also indebted to ICI (Chemicals and Polymers Group) plc, BP Chemicals (London), and Ciba-Geigy plc for the provision of information and materials, and to Viritech Ltd. (Swindon) for the use of their electron-beam irradiation facilities.

REFERENCES

1. Advisory Committee on Irradiated and Novel Foods, report on the 'Safety and Wholesomeness of Irradiated Foods', HMSO, London, April 1986.

2. D.W. Thayer, in 'Food and Packaging Interactions', Ed. J.H. Hotchkiss, A.C.S. Symposium Series 365, American Chemical Society, Washington DC, 1988, Chapter 15, p.181.

3. K. Figge and W. Freytag, Deutsche Lebensmittel-Rundschau, 1977, 73, 205.

4. D.W. Allen, J.S. Brooks and J. Unwin, Polym. Deg. and Stability, 1985, 10, 79.

5. D.W. Allen, J.S. Brooks, J. Unwin and J.D. McGuinness, Chem. Ind. (London), 1985, 524.

6. D.W. Allen, J.S. Brooks, J. Unwin and J.D. McGuinness, Appl. Organometallic Chem., 1987, 1, 311.

7. D.W. Allen, J.S. Brooks, and J. Unwin, Appl. Organometallic Chem., 1987, 1, 319.

8. P-L. Horng, and P.P. Klemchuk, Plast. Eng., 1984, 4, 35.

9. G.D. Mendenhall, H.K. Agarwal, J.M. Cooke, and T.S. Dzimianowicz, in 'Polymer Stabilisation and Degradation', Ed. P.P. Klemchuk, A.C.S. Symposium Series 280, American Chemical Society, Washington DC, 1985, p.373.

10. D.J. Carlsson, J.B. Dobbin, J.P.T. Jensen and D.M. Wiles, in 'Polymer Stabilisation and Degradation', Ed. P.P. Klemchuk, A.C.S. Symposium Series 280, American Chemical Society, Washington DC, 1985, p.359.

11. J. Brodilova, B. Chutny, and J. Pospisil, Die Angew. Makromol. Chem., 1986, 141, 161.

12. D.W. Allen, D.A. Leathard, C. Smith and J.D. McGuinness, Chem. Ind. (London), 1987, 198.

13. J. Brodlihova, B. Chutny and J. Pospisil, Die Angew. Makromol. Chem., 1986, 141, 161.

14. D.W. Allen, D.A. Leathard and C. Smith, Chem. Ind. (London), 1988, 399.

15. K. Figge, Progr. Polym. Sci., 1980, 6, 187, and references therein.

16. Commission of the European Communities, Council Directive COM (78), 115 Final, Brussells, 29 May 1978.

17. N. De Kruijf, M.A.H. Rijk, L.A. Soetardhi, and L. Rossi, Food Chem. Toxicol., 1983, 21, 187.

18. L. Gramiccioni, P. Di Prospero, M.R. Milana, S. Di Marzio, and I. Marcello, Food Chem. Toxicol., 1986, 24, 23.

19. Anon., Food, Cosmetics and Drug Packaging, 1986, 9, 12.

20. K. Figge and W. Freytag, Deutsche Lebensmittel-Rundschau, 1977, 73, 205.

21. K. Azuma, H. Tsunoda, T. Hirata, T. Ishitani, and Y. Tanaka, Agric. Biol. Chem., 1984, 48, 2009.

22. F. Lox, R. De Smet, A. Walden and J. Machiels, Proceedings, 5th International IAPRI Conference, Bristol, 7-9 October 1986.

23. D.W. Allen, A. Crowson, and D.A. Leathard, Chem. Ind. (London), 1990, 16.

24. D.W. Allen, D.A. Leathard and C. Smith, Chem. Ind. (London), 1989, 38.

Irradiation of Packaged Food

D. Kilcast
LEATHERHEAD FOOD RESEARCH ASSOCIATION, RANDALLS ROAD,
LEATHERHEAD, SURREY KT22 7RY, UK

1   INTRODUCTION

Food irradiation is used to improve the safety of food by killing insects and microorganisms, to inhibit sprouting in crops such as onions and potatoes and to control ripening in agricultural produce.

In order to prevent re-infestation and re-contamination it is essential that the food is suitably packed. Consequently, the packaging material is irradiated whilst in contact with the food, and it is important that the material is resistant to radiation-induced changes. In this paper the nature of the irradiation process is reviewed briefly, together with the known effects of irradiation on packaging materials and their implications for the effective application of food irradiation. Recent research carried out at the Leatherhead Food RA on the possibility of taint transfer into food is described.

The Irradiation Process

Food irradiation involves treatment of food with gamma rays or electron beams. These represent forms of ionising radiation and, in water, the major component of foods and living organisms, ionisation results in the formation of highly reactive free radicals and hydrogen peroxide. The amounts produced and subsequent chemical changes depend on dose. In general, at radiation doses appropriate to food (<10 kGy), the DNA molecule is the most sensitive site and is inactivated in living insects and microorganisms, causing death. Chemical changes that occur in the food itself are much less than those that occur in cooking.

Two main sources of ionising radiation are used for food irradiation - a radioactive source, usually cobalt-60, which emits gamma rays, or a linear accelerator supplying high-energy electrons.[1] Gamma rays have high penetrating power, and consequently cobalt-60 sources are used for treating bulk items such as sacks or drums of food. High-energy electrons have limited penetrating power (maximum 8 cm for two-sided irradiation with 10 MeV beam energy), but are highly suited for treatment of products such as grain moving as a thin layer on a conveyor belt, and can also be used for the treatment of trayed chilled meals.

The irradiation dose used is determined by the purpose of the irradiation treatment and by the type of food. Although a greater effect is obtained at higher doses, there are increasing risks of off-flavour development and textural softening in food. Sterilisation requires doses >10 kGy, and doses <1 kGy are used for disinfestation, ripening control and sprouting inhibition.

For most potential applications, packaging materials will therefore need to resist radiation doses of up to 10 kGy. Most published information on the effect of irradiation in packaging is, however, concerned with doses greater than this. Three specific requirements for packaging materials can be defined.[2] The materials must

i) be resistant to radiation with respect to their functional protective characteristics;

ii) not transmit toxic substances to the food;

iii) not transmit undesirable odours or flavours to food.

The Effect of Irradiation on Packaging Materials

Information on the effect of irradiation on packaging materials has been acquired from a number of sources: firstly, from the use of irradiation for the modification of the properties of plastics in the dose range 50-250 kGy, for example shrink-wrap film; secondly, from the use of irradiation for the sterilisation of medical products at doses >25 kGy; and thirdly, from work specifically aimed at investigating the use of packaging materials for use in food irradiation. In addition, gamma irradiation is also being used in the preparation of aseptic bulk bag-in-box packaging materials for food use as an alternative to thermal or hydrogen peroxide treatments.

Rigid metal containers such as tin or aluminium cans are highly resistant to radiation at doses likely to be used for foods. The physical and chemical properties of glass are unaffected at food irradiation doses, but even at relatively low doses (<3 kGy) glass and ceramic materials become brown.

Cellulosic materials such as paper and board are less resistant to radiation-induced changes, owing to changes to the polymer structure. This can in principle lead to a loss in strength, but this may have little practical consequence if the material is supported by polythene or foil as a laminate. Secondary packaging materials such as wood should be satisfactory for use in irradiation processing.

The most common form of packaging material for use in food irradiation is likely to be some form of thermoplastic film or laminate. Plastics may undergo one of two basic reactions on irradiation, depending on the nature of the materials and the irradiation conditions. Cross-linking between chemical units may occur, and may convert a two-dimensional structure into a three-dimensional structure, leading to an increase in the tensile and flexural strength of the film. Conversely, degradation of the polymer chains to shorter units may occur, leading to a reduction in overall strength and increased porosity. In addition, the irradiation treatment may interact with additives in the packaging material such as plasticisers or stabilisers and, in laminated materials, there may be an effect on the adhesive. Finally, there is a possibility that components of printing inks may be affected by irradiation. Research has shown that physical and chemical changes due to cross-linking or degradation are not significant in cellulosic materials at doses <10 kGy, and in synthetic polymers at doses <60 kGy. Changes in gas and water permeability are generally not significant at doses <8 kGy. Gas evolution can occur, however, in certain materials when irradiated at doses <10 kGy. The most common gases evolved are hydrogen and low-molecular weight hydrocarbons, and halogenated polymers, e.g. polyvinylchloride, can release hydrogen chloride.

Studies on the migration of additives from packaging materials into food on irradiation have been carried out, and on the basis of this work the FDA approved in 1964 a range of films as safe for use for gamma irradiation up to 10 kGy (Table 1).

Irradiation of Packaged Food

Foods that are irradiated for insect disinfestation and stored for long periods, for example cereals and dried fruit, require packaging to prevent recontamination. Such packaging must be sufficiently durable for the purpose.

Produce that is to be irradiated to increase shelf-life through reduction in the numbers of spoilage organisms requires packaging appropriate for that purpose. The use of non-perforated packaging for fruits and vegetables will result in modified gas atmospheres in the pack. Low oxygen and high carbon dioxide atmospheres in the packs could delay senescence in some products, further enhancing the radiation-induced shelf-life extension. Sealed packaging will also act as a water barrier, reducing weight loss, and as an oxygen barrier, reducing radiation-induced vitamin C loss.

Table 1 FDA-approved Polymeric Films to 10 kGy

Nitrocellulose or vinylidene coated cellophane
Wax coated paperboard
Glassine paper
Polypropylene film
Polystyrene
Rubber hydrochloride
Vinylidene chloride-vinyl chloride
Polyethylene
Polyethylene terephthalate
Nylon 6
Nylon 11
Vinyl chloride-vinyl acetate
Vegetable parchment

Extension of the shelf-life of fish, poultry and red meat products can be achieved using irradiation. Packaging materials should maintain their required oxygen and moisture barrier properties, and most studies have shown minimal problems, although off-odours have been detected in products packed in polyethylene and polypropylene. Problems associated with the presence of oxygen and subsequent fat oxidation and rancidity development can, in principle, be minimised by vacuum packing. More work is required on the effect of residual oxygen and of oxygen permeability of packaging films on the sensory properties of irradiated meat.

Research into irradiation sterilisation of food at doses >10 kGy has shown that flexible packaging in the form of a pre-formed multi-layer pouch consisting of nylon, aluminium foil and polyethyleneterephthalate-polyethylene is suitable for red meat and poultry.

It should be noted that the FDA list, which was drawn up in the late 1960s following research by the US Army into the production of irradiation-sterilised food packs, bears little relationship to the wide range of packaging material that would be used for likely applications in the near future. In particular, the list does not include co-extruded materials or laminates and does not refer to the use of electron beam irradiation.

Food Taints

Taint can be defined as an unpleasant flavour in foods that is present through transfer from external sources. The chemical species that give rise to taint are generally present in very low concentrations and do not present any toxicity hazard, but are characterised by low thresholds. Consequently, significant numbers of consumer complaints can result from ppm or even ppb concentrations. Identification of these very low levels when present in complex foods by chemical means (e.g. GC-MS) can be difficult, and human subjects possessing high sensitivities are often more reliable in establishing the presence of a particular tainting species. For example, many food manufacturers maintain panels of known sensitivity to chlorophenols and chloroanisoles, which are known to be the source of many food taint problems.[4]

Irradiation of Packaged Food

Taint problems have been known to occur from migration of species from packaging materials into food. The food-packaging and the food-manufacturing industries usually guard against this by the use of some form of sensory test procedure, commonly sensory difference testing.

Such tests are important in screening out packaging materials, including additives such as antioxidants and UV stabilisers, that could cause taint problems.

The limited chemical changes occurring in packaging material at food irradiation doses, although unlikely to give rise to mechanical problems or toxicity problems, carry the risk of taint problems through transfer of low levels of breakdown material. The remainder of this paper deals with experiments carried out at the Leatherhead Food RA to examine the possibility of taint transfer from common polymeric food packaging materials into food simulants during, and following, irradiation.

2 EXPERIMENTAL PROCEDURE

General

The packaging materials tested in this study are shown in Table 1. Irradiation was carried out using a cobalt-60 source at Isotron Ltd, Swindon, at a dose rate of 8.4 kGy per hour for all materials except PVC, PET and PS tray material, which were irradiated at 6 kGy per hour. Radiation doses were measured by Amber Perspex dosimetry, and average doses are shown in Table 2.

Selection of Food Simulants

Food simulants were chosen using three criteria: firstly, their ability to simulate the range of solvent properties commonly exhibited in foods; secondly, acceptable palatability for organoleptic testing; thirdly, their resistance to irradiation - induced flavour changes. Previous experiments had shown that water and ethanol solution (8% v/v in water) satisfied these criteria. More difficulty was encountered in selecting a fat simulant. All liquid oils tested (groundnut, olive, sunflower, rapeseed, corn and Durkex 500) gave unpleasant flavours following irradiation. Hydrogenated palm kernel oil (HPKO) was consequently

Table 2   Packaging materials tested and average irradiation dose

Material	Dose/kGy
50 μm polyethylene	2.6
* PVdC - coated polyester/polyethylene	3.8
Polyethylene + slip additives	3.8
Polypropylene	3.5
* PVdC - coated polypropylene	3.5
* PVdC - coated cellophane	3.5
Nitrocellulose - coated cellophane	3.5
Polyvinyl chloride (PVC) tray	3.9
PVC cling film	3.9
Polyethylene terephthalate (PET) tray	3.5
Polystyrene (PS)	3.5

*   PVdC - polyvinylidene chloride

selected on the basis of its stability and palatability.

Packaging/Simulant Preparation

Packaging films were made up into heat-sealed pouches at a ratio of 2 dm² of film to 100 ml of liquid (water and aqueous ethanol) and 2 dm² of film to 50 g of HPKO. These were made up on the day before irradiation and stored at 4°C. Simulants for control pouches were prepared the day before irradiation and stored in glass at 4°C. Non-irradiated packaging film was made up into pouches containing the irradiated simulants on the day of irradiation.

The same ratio of packaging material to water and alcohol simulants was used for the PVC, PET and PS tray

Irradiation of Packaged Food

material. In this case, the trays were filled with the simulants and sealed into pouches made from 50μm polythene film. For practical reasons, the following amounts of HPKO were used for the trays: 80 g HPKO to 240 cm² PVC, 40 g HPKO to 120 cm² PET, 40 g HPKO to 113 cm² PS.

A similar method was used for PVC cling film in contact with HPKO. Glass dishes were covered with cling film and 45 g HPKO to 133 cm² film placed inside a lid of the dish. For PVC cling film with water and aqueous ethanol simulants, a 91.4 cm x 7.62 cm strip was cut into 7.62 cm x 7.62 cm squares and immersed in 600 ml of simulant.

The HPKO systems were transported for irradiation at ambient temperature, the liquid systems at approximately 10°C. Following irradiation, the systems were stored at 4°C for 14 days.

The experimental scheme is shown in Figure 1.

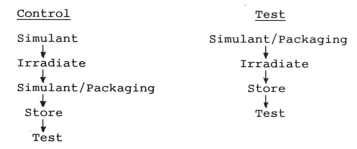

Figure 1 Experimental scheme for packaging tests

Sensory Testing Procedure

Control and test simulants were tested against each other in a triangular test (BS5929 Part I (1990) and Part II (1984)), using fifteen experienced assessors. In this test, panellists are presented with three coded samples, two of which are the same and one different, using all possible permutations (i.e. CCT, CTC, TCC, CTT, TCT, TTC, where C = control and T = test). The panellists are asked to identify the odd sample, and are also asked how confident they are in that choice, in what way the samples differ and which they prefer.

Statistical tests were carried out on the number of correct identifications of the odd sample. Significance levels of less than 20% were judged to indicate that a difference had been found between test and control samples. This high level was chosen to minimise the risk of Type II errors, i.e. of not identifying a real difference.[5] As taint is often only identified by a minority of assessors, minority judgements were examined carefully in conjunction with the ancillary information, especially preferences and descriptive terms.

3 RESULTS AND DISCUSSION

Summary results for individual packaging materials are given below, together with illustrative experimental data on individual materials.

Neither type of polyethylene showed any indication of taint transfer, confirming earlier results. However, the PVdC-coated polyester/polyethylene laminate showed some evidence for taint when the water simulant was used. Other PVdC-coated materials did not show any consistent evidence for taint, indicating that the lamination adhesive might be the source of taint for the laminate.

No evidence for taint transfer was found for polypropylene. Nitrocellulose-coated cellophane showed evidence for taint transfer into both water and alcohol simulant, confirming earlier data. However, this material is unlikely to find a use with irradiated foods.

There was evidence for taint transfer from both PVC tray material (into alcohol simulant) and PVC cling film (water simulant). These data are shown in Tables 3 and 4, respectively.

In these and other tables, the following key should be used:

SL = significance level
CC = correct choice of cold sample
IC = incorrect choice of cold sample
ND = no difference judgement
C = control
T = test
NP = no preference

Descriptions recorded for the irradiated alcohol/PVC tray system included strong, aftertaste, bitter/strong. Descriptions recorded for the water/PVC cling film system included cardboard, musty, chlorinated, metallic, tainted and horrible. The PVC cling film tested was standard overwrap for fruit and vegetables, and not low-migration material. Earlier tests using domestic grade cling film had shown taint in both alcohol and HPKO simulants, but not in water.

Table 3 Test Data for PVC Tray Material

	Test responses			SL	Preferences		
	CC	IC	ND	%	C	T	NP
water	4	3	8	60	2	2	0
alcohol	9	0	6	3	8	0	1
HPKO	3	2	10	86	1	1	1

Table 4 Test Data for PVC Tray Material

	Test responses			SL	Preferences		
	CC	IC	ND	%	C	T	NP
Water	12	2	1	0.01	9	0	3
alcohol	6	1	8	32	5	0	1
HPKO	3	0	12	92	1	0	2

Table 5   Test Data for PS Tray Material

	Test response			SL	Preferences		
	CC	IC	ND	%	C	T	NP
water	4	4	7	53	0	1	3
alcohol	10	0	5	0.85	5	1	4
HPKO	1	1	13	99	0	1	0

PET tray material showed some indication of taint transfer into the water simulant, but the evidence for this was weak.

PS tray material showed evidence for taint in the alcohol simulant (Table 5).

Descriptions recorded included plastic, musty, choking, strong and tainted. The high-impact polystyrene tested in this study is generally used for margarine tubs, yoghurt cartons, etc. Expanded polystyrene is more likely to be used in contact with irradiated foods.

In the case of polyethylene- and polypropylene-based films, there was some evidence that the non-irradiated materials tainted the simulants to a mild degree, but that this effect was removed on irradiation. This may be a result of the trapping of mobile species as a result of irradiation - induced free radical reactions.

In determining the likely practical relevance of the results from this study, it should be noted that the tests and their interpretation were designed to be highly critical of packaging material performance. This is a direct consequence of the commercial consequences of not identifying a potential taint problem before tainted food reaches consumers. A taint complaint rate of 1 in 1,000 can be sufficient to generate major problems for retailers and manufacturers, and financial losses based not only on the value of the tainted stock but also from litigation and from damage to a brand name. Identification of potential taint problems using sensory methods consequently demands the severe testing conditions used in this work, particularly when the chemical nature of tainting species is unknown.

Of the materials tested in this study, PVC carries a clear risk of tainting food on irradiation, and does not appear on the FDA approved list. In view of its commercial importance, particularly as an overwrap, steps should consequently be taken to avoid the use of PVC for irradiation. The position with other materials is less clear, but as other important material such as polystyrene may carry some risk of taint, tests should be carried out using any foods likely to be irradiated in contact with this material.

The observation that irradiation removed evidence for taint transfer from polyethylene and polypropylene-based films is in accord with previous observations that the migration of antioxidants into food simulants can be reduced on irradiation.[6] It must be stressed, however, that a combination of both sensory testing and chemical analysis would be needed to investigate fully the relationship between migration and taint.

In view of the extensive current use of new types of polymeric packaging materials, including co-extrudates and laminates, more screening is needed on these materials in order to examine their suitability for irradiation. In addition, additives, adhesives and printing materials should also be screened. Some work has also indicated that the extent of migration also depends on whether gamma or electron beam irradiation is used.[7] This may be a result of either the nature of the radiation or, more likely, the faster dose rate delivered by an electron beam (up to 100 times greater than gamma rays). Both of these potentially important sources of radiation should consequently be examined.

REFERENCES

1. W.M. Urbain, 'Food Irradiation', Academic Press, London, 1986, Chapter 15.

2. S.R. Agarwal and A. Sreenivasan, J. Fd Technol., 1972, $\underline{8}$, 27.

3. D.W. Thayer, 'Food and Packaging Interactions', ed. T.H. Hotchkiss, American Chemical Society, Washington, 1986, Chapter 15.

4. F.B. Whitfield, 'Developments in Food Flavours', ed. G.G. Birch and M.G. Lindley, Elsevier, London, 1986, Chapter 15.

5. M.J. O'Mahoney, 'Sensory Evaluation of Food : Statistical Methods and Procedures', Marcel Dekker, New York, 1986, Appendix F, p 401.

6. D.W. Allen, D.A. Leathard and C. Smith, <u>Chem. Ind.</u>, 1988, 399.

7. F. Lox, R. De Smet, A. Walden and J. Machiels, Proceedings, 5th International IAPRI Conference, Bristol, 7-9 October, 1986.

Commercial Food Irradiation in Practice

J. G. Leemhorst

GAMMASTER INTERNATIONAL B.V. EDE, THE NETHERLANDS

1 INTRODUCTION

For outsiders The Netherlands is not only famous for its water-works, but also for its high standard of farming and food production. The yield of our small surface of arable land is so high, that the surplus in production makes us the third largest exporter of agricultural and horticultural products in the world.

This advanced position has been achieved through a continuous education and training effort for workers in food production and through the development and implementation of technologies that could improve production and/or quality.

It is no wonder that Dutch research showed great interest in the potential of food irradiation at an early stage. The positive research results and the potential applications for industry encouraged the Ministry of Agriculture and Fisheries to construct a Pilot Plant for Food Irradiation.

2 THE PILOT PLANT FOR FOOD IRRADIATION

In 1967 the Pilot Plant for Food Irradiation in Wageningen came into operation. The project was financed by the Ministry of Agriculture and Fisheries and several commodity boards. Scientific back-up was secured through the Institute for the Application of Nuclear Technology in Agriculture. This institute was related to Eurisotoop and the Agricultural University at Wageningen and specialised in nuclear research in relation to agriculture.

The objectives of the Pilot Plant were:

- Research into the practical application of irradiation on agricultural and horticultural products,
- Introduction of the irradiation technology into the food industry,
- Efficacy testing and test marketing of irradiated products,
- Information transfer to the public.

The legal basis of the activities of the Pilot Plant was created through the introduction of a positive clearance policy, gradually permitting irradiation of groups of products after thorough investigation. For each product (group) the Pilot Plant wanted to investigate, a petition with details on the product, the reasons and the process to be applied had to be sent to the Minister of Health. The Minister of Health consulted an expert committee and depending on the judgement of the experts about possible health hazards, was able to grant clearance in three categories.

I   not for trade purposes, maximum quantity 1,000 kg.
II restricted quantities, mostly a few tons destined for test marketing.
III unrestricted quantities with an unconditional clearance.

For a long time this positive clearance system through the intervention of an expert committee has been unique. The liberal attitude of our government together with the activities of the Pilot Plant and later the irradiation industry has given The Netherlands its advanced position in the number of products that are cleared for irradiation.

The Pilot Plant was equipped with a Van de Graaff-electron-Accelerator and an automatic gamma-irradiator. Especially in the first 6 years of operation many activities were developed. While the financing was taken care of by the agricultural sector, test irradiation and marketing of fresh products and fruits was emphasised. Later other products were investigated. Much attention was paid to information transfer to the public. All media channels, government organisations, consumer organisations and teaching programmes in food technology have been and still are involved in information transfer. Recent polls showed that about 50% of the Dutch population is aware of the application of the process to foods.

3 ACCEPTANCE BY INDUSTRY

In spite of the many efforts made by the Pilot Plant and a rational reception by the public and the food industry, industrial application of the process started only 11 years after the introduction. The main reasons for this slow implementation are:

- The change in appreciation of nuclear technologies in the seventies,
- The products focused on by the Pilot Plant,
- The restrictions of some other importing countries on the trade in irradiated foods.

Especially the latter two reasons are interconnected. The extension of the shelf-life of strawberries, asparagus, mushrooms and endives were shown to be successful applications of irradiation. The fact that the domestic consumption is only a small fraction of the total overproduction and that our main customer, The Federal Republic of Germany, has a ban on irradiated foods, made the application to these products not economically justified.

Irradiation of tubers for sprout inhibition was shown to be a very effective process, but required transportation to a central irradiation facility and additional handling. As long as easy applicable, low cost chemicals are allowed for this purpose irradiation will be too expensive.

The irradiation of poultry for the prevention of salmonella and campylobacter contamination was a very promising application. The lack of regulations regarding the permissible levels of contaminants and the strong competition in the poultry industry prevented the application of the process for a long time.

Forced by diminishing funds, the Pilot Plant started after the mid-seventies to use the automatic gamma-irradiator for the irradiation of other products. The decontamination of shrimps, spices and herbs showed to be economically feasible and to fulfil a market demand. The demand soon became so high that the small capacity of the Pilot Plant was not sufficient and in 1978 help from the commercial irradiation company Gammaster was sought.

Gammaster, a subsidiary of the Dutch cooperative pharmacist organisation OPG UA, was operating an automatic gamma-irradiator in Ede near Wageningen and specialised in the GAMMA-STERilisation of medical supplies.

4 THE INDUSTRIAL GAMMA-IRRADIATION FACILITY

Most of the approximately 160 industrial gamma-irradiators, operating all over the world, are used for the sterilisation of medical supplies produced in mass production. The efficient processing of large volumes of similar low density products, packed in standard boxes and treated with a standard high irradiation dose, has always dictated the design. Optimal use of the available radiation (high radiation efficiency) and good dose uniformity had priority above flexibility and product handling.

The automatic tote box irradiator, in use since 1971, at Gammaster in Ede is a typical medical-irradiator. The transport system only accepts boxes of uniform dimensions (48 x 61 x 90 cm) and the source is completely surrounded by the product, causing a large product hold-up in the irradiation room.

A shuffle-dwell-system is used for the transport through the irradiation cycle. This means that when one product box enters the irradiation cycle all the boxes in the cycle are moved one position ahead. At the end of the movement the product box which has passed all irradiation positions leaves the cycle and everything stays in rest until the next product box enters the cycle. The time interval between the entrance of two product units counts for the total time a box will remain in the irradiation cycle. The dose received by the product is directly related to the irradiation time and can only be influenced through a variation of the time interval.

This system meets the requirements very well in a standardised process but causes high inefficiencies when products with different densities have to be irradiated or when different doses have to be applied. To ensure that all boxes in the cycle receive the right dose, first the total irradiation cycle has to be emptied before the time interval can be altered. In the Ede situation 75 boxes have to be transported into the system before the time interval can be changed. The time needed when changing from high dose medical sterilisation to another process is, even at 1 million Curie source loading, about seven hours.

When we started to use this facility for the low dose irradiation of high density foods we were confronted with these inefficiencies and several other problems. The transport-system, designed for the application of a minimum 25 kGy sterilisation dose, could not transport the product fast enough through the irradiation cycle. Even after modifications to speed up the system, the lowest applicable dose at a 800 kCi source was about 3 kGy. Most of the food packagings did not comply with the standard box dimensions and much handling was required to pack the products in acceptable boxes.

5 THE PALLET IRRADIATOR

As a commercial irradiator, only performing contract irradiation for other companies, Gammaster had to be inventive in order to sell its services. As early as 1974 the irradiation with 8 kGy of laboratory utensils and packaging for liquids was introduced by Gammaster. The renewed interest in food irradiation was the incentive for a market investigation and the automatic irradiator was used for testing.

The positive results of the market investigation and especially the indication that the FRG intended to lift the ban on the irradiation of spices and herbs, supported the decision to become active in this field too.

This decision and the growing demand for services in the other application areas led Gammaster to extend the operation by building an additional irradiation facility in 1979. Based on the experience with the tote box irradiator, a completely different concept was needed.

A real multipurpose irradiator should meet the following criteria:

1) Labour involvements should be minimal.
2) No limitation should be given by product dimensions.
3) Products with densities up to 0.5 g cm^{-3} should be acceptable.

4) It should be possible to apply any requested dose from 0.05 kGy upwards.
5) The dose uniformity at low densities should be comparable with the uniformity of our tote box irradiator.
6) Inefficiencies related to the use of empty boxes, when changing over to another dose setting should be avoided.
7) At least 40 hours unattended operation of the unit should be possible.
8) The source rack should have enough capacity for 15 years operation with a 3 Mega Curie source.

After long discussions, we were eventually able to convince Atomic Energy of Canada Limited to construct a facility that incorporated these demands. This completely new irradiator came into operation in 1982.

In this unit large carriers, hanging on an overhead-monorail system, are used for product transport. Each carrier can hold two pallets with dimensions of 100 x 120 x 180 cm. The total product volume per carrier is $2 \times 2.16 \, m^3 = 4.32 \, m^3$. The transport-system can handle one carrier every two minutes, which equals a total process capacity of 130 m^3 an hour.

The maximum source capacity is 3 Million Curies of Cobalt-60, which, even with the fast transport-system, could create a limitation on the lowest dose to be applied. Therefore, a special cylindrical source rack, consisting of two independent cylinders, had to be designed. Depending on the dose to be applied it is possible to bring one or both cylinders into the irradiation position. To improve dose uniformity this source, with its length of 4.50 m, overlaps the product. This sacrifices radiation efficiency but avoids product lifting in the radiation room, a fault sensitive procedure. When not in use, this source is stored in a 9 m deep water basin.

6 THE INCREMENTAL DOSE SYSTEM

To minimise inefficiencies caused by the use of empty carriers, when changing over from one dose to another, we extended the operation with a computer-control-system. Using this system it is possible to programme per carrier the number of cycles to be made around the source; so different doses can be given to each individual carrier, on condition that the final dose is a multiple of the dose received during one cycle. The computer-control-system also generates a process history file containing all important process-data for each carrier. Similar to the practice in medical processing all process records are kept for a period of 5 years.

7 INTRODUCTION OF COMMERCIAL FOOD IRRADIATION

For the introduction of commercial food irradiation Gammaster developed a special marketing strategy. The main components of this strategy were an "OPEN DOOR" policy and a high degree of service. Before the opening of the pallet irradiator the symposium 'Food Irradiation Now' was organised. This symposium intended for information transfer to industry and officials active in food inspections, was very successful and attracted much attention, also internationally.

When the pallet irradiator came into operation the national and international media and food sector magazines paid much attention to the new concept.

Interest was shown by many organisations from all over the world and in addition to the many national visitors more than 2,000 foreign visitors were shown around in the first 3 years of operation.

For contacts and guidance in the food industry, a food engineer was taken on and special programmes were developed for technical and feasibility studies of possible applications. An intensive advertising campaign and participation in food exhibitions and congresses supported the introduction. A recent poll showed that about 85% of the Dutch industrial food technologists are aware of our activities. That almost 50% asked for more information, forces us to continue with this information transfer.

In the 6 years of operation of the pallet irradiator we learned a lot. Labelling obligations and export restrictions are still the major obstacles for industrial acceptance. Several promising applications failed for these and for technological reasons. We experienced that positive laboratory results are not a guarantee for successful large scale application.

Depending on the origin and other factors, variations in the properties of similar products can influence the applicable process-parameters.

Over the years an extensive library has been built up and protocols for the treatment of different products have been established. A second food engineer has been hired for quality assurance and a system of Good Irradiation Practice has been implemented. As a pioneer in the field we have to develop most procedures and standards ourselves and much of our experience in medical sterilisation has been adapted to this application.

In 1983 we started with the operation of our second pallet irradiator in Munich, FRG. This unit is mainly used for the treatment of non-foods, while the ban on food irradiation has still not been lifted but occasionally foods for export and animal fodder are irradiated.

The throughput in human foods in Ede has grown to 400 tons weekly. In addition to shrimp and poultry mostly ingredients are treated. Main products are: spices, herbs, caseins, skimmed milk powder, dried vegetables, egg powder, soya flour, starch, animal blood plasma, minced dried vegetables, rice, and butter mix.

There is no demand for the irradiation of fresh agricultural and horticultural products. Treatment for the domestic market is not economically justified and export markets are closed by the ban on the trade of irradiated products.

8 PRACTICAL ASPECTS OF FOOD IRRADIATION

The practice in irradiation sterilisation to apply a standard minimum dose has been derived from the premise that in a GMP production the contamination level will be low and will only vary within a small bandwidth. By applying an overkill dose (mostly 25 kGy), certainty is given that the product is sterile after treatment. This is defined as 'the maximum occurrence of one viable microorganism in 1,000,000 items'.

Decontamination of food is similar to pasteurisation, the objective is the elimination or reduction of certain microorganisms. The difference is that in food irradiation the process may not impair the properties of the product. This principle, which is reflected in the permissible maximum dose mentioned in clearances for food irradiation, is of great influence in the application of the process.

The sensitivity to radiation varies from product to product. Some products already show changes in colour, odour or organoleptic properties at low doses, whereas others can stand a high dose treatment without any noticeable change. The threshold dose at which changes occur, has been established for many food products. But factors such as temperature, water content, age, etc can cause great variations. It is therefore necessary that the selection of process conditions and process-parameters is based on a thorough knowledge of the product and its packing.

Before starting the irradiation of a new product or treatment for a new customer, a series of steps has to be taken. Almost from the beginning the food engineer is involved. He will contact his counterpart from the manufacturer and familiarise himself with all aspects of the product. Together they will clearly define the objective of the treatment eg:

- if shelf-life extension is requested, how many days, in which storage and transport conditions,
- if reduction of the microbiological load is the objective, which organisms have to be reduced, to what level, etc.

Based on the extensive information in our library and our expertise with similar products, already at this stage an indication can be given of the feasibility of the objective.

When the available information is not sufficient a series of tests at different doses and different conditions is conducted. Before and after treatment the product is investigated by an external microbiological laboratory and the reduction of the microbiological load is established.

In general the reduction of the following organisms is of the interest:

- Total plate count g^{-1}
- Enterobacteriaceae g^{-1}
- <u>Staphylococcus aureus</u> g^{-1}
- Yeasts and moulds g^{-1}

If requested the D_{10} dose, that is the dose that gives a reduction of 90% of a certain organism, is established by the application of doses increased in stages. Once the D_{10} value is known, the decision can be made as to which dose has to be applied to have the required reduction. Parallel to the microbiological investigations, the influence of the different doses on the properties of the product and packing is established.

This is usually done by the manufacturer, but use is also made of specialised institutes. It is important to know, whether the threshold dose of impairment is below or near the requested reduction dose. If this dose is below, the application of a combination process eg irradiation at low temperature or in adapted environment can be considered.

When the requested dose is below but near to the impairment dose, special attention has to be paid to the process conditions. There is always, depending on the density and dimensions of the product and the specifications of the irradiation facility, a variation in the dose received at different positions within the product unit. By adaptation of the process-parameters it has to be made certain that the maximum dose received is not above the impairment dose and the minimum dose is not below the required reduction dose.

When the tests show positive results and the decision is made to irradiate large volumes, often a dose-mapping is performed. Dosimeters are placed on positions throughout the product unit and the dose distribution in the product is measured. With this procedure the maximum and minimum dose positions are located. These are the positions at which in routine treatment the control dosimeters are placed.

As a result of 8 years operation of the pallet irradiator, the dose distribution curves are well known for many densities and product presentations. Corrections have to be made when the source configuration is changed (eg source replenishment). This expertise and the knowledge about the radiation influence on the product enables the food engineer to advise the customer on the optimal presentation of the products to the irradiation process.

9 QUALITY CONTROL

The Codex General Standard for Irradiated Foods states:

The irradiation of food is justified only when it fulfils a technological need or where it serves a food hygiene purpose and should not be used as a substitute for food manufacturing practices.

It will be clear that in industrial application, compliance with the first part of this sentence is ruled by economic considerations. No manufacturer will incur additional costs when it does not serve a real purpose. The second part of the sentence is very general and a cause of concern for those not familiar with the process. It is often interpreted to mean that irradiation can be used:

- to mask bad quality,
- to improve spoiled foods,
- to suppress spoilage indicators like odour and taste,
- to relax GMP regulations.

Those familiar with the process will know that this concern is not justified. Irradiation cannot reverse spoiling processes. It can delay microbiological spoilage but, at doses applied in food irradiation, will have almost no influence on the enzymatic processes. Investigations with the Howard-mould-test show the same findings before and after irradiation. Neither will the process remove nor modify any substance from the product. When the product is bad, while GMP is not applied, it will be just as bad after the irradiation process. So quality control in food irradiation is not different from quality control in other food treatments.

Quality control starts at the receipt of the product, when the condition of product and packing is judged, and the details of importance are noted on the accompanying papers. Sometimes product in bad condition is refused or when damaged during transport photographs are taken and in consultation with the owner return or repacking is arranged. Depending on the product, temperature measurement is done and samples are taken for microbiological control. The quality assurance manager may even delay the time of treatment until the laboratory report has been received.

The conditions of acceptance, storage, handling, processing and release may vary from product to product. In addition to the standard inspection points during the flow through the facility, our food engineer in charge of quality assurance develops protocols per product group. In these protocol procedures, responsibilities and action points will be clearly defined. For this quality assurance programme we refer to the well established FDA inspection procedures which are applied in our medical sterilisation process.

On release of the product the customer receives a copy of the laboratory report and on request an irradiation certificate with the main process data. The automatically produced process history record and copies of the laboratory report and certificate are filed for a period of 5 years.

10 ECONOMICS OF MULTIPURPOSE OPERATION

The low labour requirement of the pallet irradiator is certainly important in the control of the variable costs. Of equal interest is the computer controlled processing. Through the application of the incremental dose system, which makes it possible to programme the number of laps around the source for each individual carrier, inefficiencies when changing over from one dose to another are minimised. The incremental dose system enables the application of different irradiation doses per carrier, on condition that the requested dose is a multiple of the dose received in one lap.

The permission for unattended operation outside normal working hours is also of economic importance. In the daytime carriers are loaded and processing continues during the night. By performing high dose applications at night and in the week-end up to 150 effective operation hours a week are realised with only one workers shift. Any deviation in the processing with unattended operation causes an immediate interruption of the process. The source is lowered to the safe storage position and the technician on duty is called by an automatic alarm.

The features of unattended operation, the possibilities to mix products for different dose applications and the application of high dose treatment outside normal working hours makes the multipurpose operation highly efficient and keeps processing costs low.

Commercial Food Irradiation in Practice 165

The costs of irradiation depend on:

- the dose to be applied,
- the density of the product,
- the acceptable bandwidth of dose variation,
- the handling,
- the conditions of the process.

Two examples of tentative prices including routine dosimetry and routine microbiological investigation are:

1. Decontamination of spices. A pallet load of 864 kg, density 0.4 g cm^{-3}, treated with 8 kGy to kill microorganisms will cost Dfl. 275, --. The price per kg is then Dfl. 0,32.

2. Pasteurisation of chicken on pallets with a weight of 1,000 kg, density 0.5 g cm^{-3}, treated with 2 kGy for <u>Salmonella</u> elimination costs Dfl. 140,-- per pallet.
The price per kg is then Dfl. 0,14.

11 THE OUTLOOK FOR IRRADIATION

When in 1983 the Codex Alimentarius Commission in their general assembly accepted a General Standard for Irradiated Foods and a Code of Practice for the operation of irradiation facilities for the treatment of foods, the expectation was that adaptation of national standards by the UN member states would be a formality. The outcome however was completely different. The attempts of several governments to furnish a legal basis for the application of the process initiated severe protest actions of oppositional groups.

By using every argument and means they could think of, by stirring up emotional fears and creating an atmosphere of criminal behaviour around the application of the process they were able to stop or hamper most developments. Officials of all branches involved became very innovative in proposing the most stringent requirements and controls and politicians became afraid to show any positive attitude.

Especially the developments in the European Community are an example of stepwise decrease of possible applications. The main obstacle is the fact that the FRG by a law made in 1957 has a ban on the applications of the process and the trade in irradiated foods. Several attempts of the Federal government to grant even exceptional clearances failed by the unwillingness of county governments to execute the proposed strict controls and in the international scene the FRG is still the main opponent. As it is custom in the EEC to compromise it can be seen that the positioning of the FRG has a disastrous effect on the developments. Starting with the Scientific Committee of the EEC that accepted the recommendations of the Joint Expert Committee on Food Irradiation but recommended only to start with a clearance for 11 foodgroups we see a downwards development in the process that should lead to harmonisation.

Under influence of German rapporteurs, coloring their reports with misleading information, the European Parliament positioned itself against the application of the process. The European Commission in turn proposed a reduction to 8 product groups to accommodate the European Parliament and it is very well possible that a further decrease of the number of foodgroups will be negotiated.

What we see is that without any valid reason the application of food irradiation is more and more restricted and subjected to requirements unequalled by those applicable to any other food process. This development affects highly the economics of food irradiation and could result in a situation that a very beneficial technology is sacrificed to emotional rejection and political exploitation.

One could put the case that by this the consumer is denied better quality and in several cases safer foods.

BIBLIOGRAPHY

'Handbook for conducting feasibility studies', IFFIT, 1986.

J.G. Leemhorst, Voedingsmiddelentechnologie, 1984, 13.

'Marketing and consumer acceptance of irradiated foods', IAEA-TECDOC-290, International Atomic Energy Agency, Vienna, 1983.

'Codex General Standard for Irradiated Foods and recommended international Code of Practice for the operation of irradiation facilities used for the treatment of foods'. Codex Alimentarius, 1984, 15.

J.H.F. Mohlmann. Radiat. Phys. Chem., 1990, 35, 795.

Radiat. Phys. Chem., 1990, 35, Nos 1-3 and 4-6.

Subject Index

Absorbed dose, 110
  units, 110
Acceptance of irradiation, 155
Additives
  avoidance, 6
  in food contact polymers, 124
  in packaging material, 142
Advisory Committee on Irradiated and Novel Foods, 124
  recommendations, 124
Alanine/ESR dosimeter, 118, 119
  usable range, 119
Analysis of nonpolar volatiles, 14
Antioxidant
  arylphosphite, 126
    extractable degradation products, 130
  degradation, 133, 134
    products, 125, 126, 129, 136, 137
  destruction on γ-irradiation, 126
  hindered phenol, 126
    extractable degradation products, 130
  labelled, 129, 131
  migration, 131-133, 135
Applications of food irradiation, 27
Arylphosphite antioxidants, *see* Antioxidants
Average dose, 114

Bacteria (*see also* Microorganisms, Spores)
  sensitivity to irradiation, 97, 103
  and growth phase, 103

Bacterial DNA, 98
Bacterial spores, 98
Bacterial survival curve, 101
  mixed population, 104
  types, 102
Blind tests, 16, 19, 40, 43
  results, 23
Bone
  crystallinity ratio, 85, 86, 88
  fragments, 90

Calibration curves
  beef, 21
  chicken, 20
  pork, 22
Cellulosic materials, 142
Ceric/cerous dosimeter, 117, 118
  reactions in, 117, 118
Changes in packaging material, 145
Chemiluminescence, 33
Chernobyl disaster, 3
Chromosome damage, 97, 98, 102
  and survival, 103
<u>Clostridium botulinum</u>, 5, 6
  survival, 105
  toxin, 105

Cobalt-60 source, 107
Combination treatments, 7, 8
Commercial food irradiation, 153
  introduction, 159
Computer control system, 159
Consumer acceptance, 8, 9, 27
Consumer concern, 29
Consumer questions, 2
Criteria for multi purpose irradiator, 157, 158

Decontamination
  of food, 160, 161
  of herbs and spices, 155
Degradation of polymer additives, 125
Delayed luminescence, 32
Delay of ripening and sprouting, 6, 7
Depth-dose curve, 111-113
Depth-dose effects, 112
Detection (see also Irradiated food)
  of irradiated meats, 24
  test, 3, 10, 25, 31
    criteria, 10, 11, 30, 80
Dichromate dosimeter, 118
  linearity, 118
  reactions in, 118
Dihydrothymidine production, 63, 77
DNA
  alterations, 98, 99
  base damage, 100
  base labelling, 73
  base modifications
    non-specific, 72
    specific, 73
  chain scission, 70-72
  damage, 58, 100
    detection, 66
    direct effects, 63
    indirect effects, 59
    in irradiated calf thymus, 70
    in irradiated wheat, 70
  double strand breaks, 100
  in foods, 57
  free radical attack, 60
  modified base analysis, 73, 74
  overall disruption, 67
  post-labelling
    assay, 73
    technique, 71
  reaction with formaldehyde, 67-69
  repair capacity, 106
  sedimentation velocity, 99
  strand breaks, 63, 99, 100
  structure, 58, 59

Dose
  distribution, 113, 114
  estimation, 23
  to kill a single cell, 102
Dose-response relationship, 13, 15, 17, 92
Dosimeter, see individual types
Dosimeter calibration, 112
Dosimetry, 9, 10, 109
  cavity effects, 112
Dyed plastic dosimeters, 120

Electromagnetic spectrum, 2
ELISA, 75, 76
  detection of dihydrothymidine, 76, 77
  detection of thymidine, 76
Energy
  absorption, 111
  stimulated release, 32, 33
ESR signal
  and bone crystallinity, 85
  and bone fragments, 90
  and bone site, 86-88
  and chicken age, 85, 86
  of chicken bone, 82
  effect of cooking, 87-89
    with post-irradiation, 89
    with pre-irradiation, 89
  of food containing bone, 82
  of fruits and vegetables, 93, 94
  and irradiation dose, 83, 84
  quantification, 90
  of mussel shell, 92
  and sample preparation, 83, 84
  and sample storage conditions, 84, 85, 92
  and shells, 90
  of shrimp shell, 91, 92
  of strawberries, 93
ESR spectroscopy, 80, 81

Fatty acid composition of meat lipids, 16
Food hygiene, 5
Food irradiation, 1, 109, 140
  and the EEC, 166
  pilot plant, 153
  practical aspects, 160

Subject Index

Food irradiation (continued)
 purpose, 4
 technical merits, 29
Food packaging, 140
Food safety improvement, 4, 5
Food simulant media, 130-132, 145
 packaging of, 146, 147
Food taints, 144
 assessment by panel, 144
Fricke dosimeter, 116, 117
 linearity range, 117
 range extension, 117
 reactions in, 116, 117

Gamma irradiation
 and antioxidant migration, 131, 132
 effects on antioxidants, 127, 128
Good radiation practices, 28
Gram-negative organisms, 103
 (see also Bacteria)
Gram-positive organisms, 103
 (see also Bacteria)

Hazard evaluation, 105
High energy electrons, 112
Hindered phenol antioxidants, see Antioxidants
Hydrocarbons from irradiated meats, 14
 standard curves, 16
8-Hydroxy-guanosine production, 73

Immunological assays, 74
Incremental dose system, 158, 159
Induction of mutations, 105
Industrial irradiation facility, 156
Inorganic material, see Thermoluminescence
Internal standards
 for antioxidant analysis, 135, 137
 stability of, 137

Ionisation in biological systems, 60
 effects, 25
Ionising radiation, 57, 97
Irradiated food
 biological changes, 26
 codex general standard, 163
 containing bone, 81
 paramagnetic species in, 81
 qualitative detection, 80, 82, 94, (see also Detection)
 quantitative tests, 30, 31
Irradiation
 certificate, 164
 costs, 165
 dose
  and application, 141
  D_{10} definition, 162
  distribution, 162
  mapping, 162
  quantification, 94
 of food-contact plastics, 124, 146
 introduction and licensing, 8
 of meat samples, 14
 of mixed micro-organism population, 105
 opposition to, 165
 outlook, 165
 of packaged food, 143
 pilot plant
  equipment, 154
  legal basis, 154
  objectives, 154
 of polymer stabilizing additives, 124
 process, 3, 140
 source, 2, 109, 132, 141, 158
  and antioxidant degradation, 133, 134
  and antioxidant migration, 133, 135
Irradiator, (see also Pallet irradiator)
 economics, 164
 rigid containers, 142
 transport system, 156, 158
 and dose, 156

Irradiator (continued)
  inefficiencies, 156, 157
  unattended operation, 164

Key hydrocarbons, 13

Labelling, 9
Luminescence detection, 25
  prospects, 52
  sensitivity and specificity, 33
  techniques, 52
Lyoluminescence, 33

Maximum dose proposals, 29
Microbial toxin, 5
Microbiological effects, 26
Microbiological flora changes, 6
Microorganisms, 97 (*see also* Bacteria)
  comparative resistance, 103
  damage resistance, 97
  loss of viability, 102
  selection of radiation resistant strains, 104
Migration
  of additives, 143
  of hindered phenol antioxidants, 130
  and taint, 151
Mutants of wild type bacteria, 98

National regulation, 28
  EC position, 28
  UK position, 28
National standardising chain, 115
New product irradiation, 161
  microbiological investigation, 161

Obstacles for industrial acceptance, 159
"Open door" policy, 159
Optichromic dosimeter, 121
Organotin stabilizers, 125, 126
  degradation and dose, 125, 126
Overdose ratio, 113, 114

Packaging, 7
  and irradiation sterilisation, 144
  material
    effect of irradiation, 141
    requirements, 141
Pallet irradiator, 157 (*see also* Irradiator)
  extension of operation, 157
  second, 160
    main products irradiated, 160
Peroxyl radicals, 63
Photons
  energy transfer processes, 110
  isoelectric scattering, 110
  production of secondary electrons, 110
Photostimulated luminescence, 51
  specificity, 50
  spectrometer, 48, 50
  spectrum, 51, 52
Poisson distribution and bacterial survival, 101
Polymer bound degradation products, 129
Polymer films
  FDA approved, 143
  future applications, 144
Practical relevance of packaging studies, 150
Prevention of bacterial attack, 155
Primary standards, 115
Procedures and standards, 160
  development, 160
Process control, 9
Process safety, 4
Product sensitivity to irradiation, 161
Prospects of developing tests, 31

Qualitative tests, *see* Irradiated foods
Quality assurance procedures, 164
Quality control, 163
Quantitative tests, *see* Irradiated foods

Subject Index

Radiation processing, 109
Radiation-resistant mutants, 106
  isolation, 106
Radiation-resistant population
  development, 106
Radiochromic dye film dosimeters, 119, 121
Radiolytic products of lipids, 13
Reactions in irradiated plastics, 142
Reference dosimeters, 115, 116
Regulatory frameworks, 27, 29
Re-irradiation, 43, 44, 49, 89, 90
Routine dosimeters, 116, 119
  response, 121, 122
  and dose rate, 122
  and fading, 122
Routine dosimeters
  and humidity, 121
  and temperature, 121

Sensory difference testing, 145
Sensory test procedures, 147
Separated minerals, 41, 44
Serious food borne pathogens, 104
Shelf-life extension, 6, 143, 144, 155
Shrimp shell, 92
Spores, 103
Sprout inhibition, 155
Statistical analysis of results, 15, 18, 19, 148

Taint, 145
  descriptions, 149, 150
  and irradiation source, 151
  from nitrocellulose coated cellophane, 148
  from polyethylene, 148
  from polypropylene, 148
  from polystyrene, 151
  potential problems, 150
  from PVC, 148, 151
  from PVdC coated laminate, 148
Thermoluminescence, 33, 34
  of adhering minerals, 34
  applicability, 49

Thermoluminescence (continued)
  discriminating power, 39
  glow curve, 35-38, 40-42, 44, 45
  of herbs and spices, 34, 35
  improved performance, 41-43
  interlaboratory trials, 40
  optical bleaching, 46
  origin of signal, 34
  sample preparation, 36
  sensitivity histogram, 39
  signal loss, 45-47
  signal stability, 40, 45, 46
  signal strength, 36
  test conditions, 36
Thymidine
  radiation products, 62
  radical anion, 64
  radical cation, 64, 66
Thymidine (continued)
  reaction with solvated electron, 64
  reactions with H radical, 65
  reactions with OH radical, 61
Thymidine glycol
  fluorimetric assay, 74
  immunoassay, 76
  production, 62-64, 74, 77
Thymine radiolysis products, 100

UK legislation, 107
Undyed plastic dosimeters, 120
UV induced damage, 107
UV spectrum
  of denatured DNA, 68
  and DNA damage, 69
  of DNA-formaldehyde complex, 68
  of native DNA, 68
Water as a reference material, 110
Water derived radicals, 60
Wholesomeness and safety, 26